Challenges of Decoding Data in Spectroscopy, Reflectometry, X-Ray and Electron Diffraction

Challenges of Decoding Data in Spectroscopy, Reflectometry, X-Ray and Electron Diffraction

By

Felix N. Chukhovskii, Petr V. Konarev and Vladimir V. Volkov

Cambridge Scholars Publishing

Challenges of Decoding Data in Spectroscopy, Reflectometry, X-Ray and Electron Diffraction

By Felix N. Chukhovskii, Petr V. Konarev and Vladimir V. Volkov

This book first published 2022

Cambridge Scholars Publishing

Lady Stephenson Library, Newcastle upon Tyne, NE6 2PA, UK

British Library Cataloguing in Publication Data
A catalogue record for this book is available from the British Library

ISBN (10): 1-5275-8604-9
ISBN (13): 978-1-5275-8604-8

TABLE OF CONTENTS

CHAPTER ONE

THE INFRARED MULTICOMPONENT SPECTROSCOPY

1.1. Introduction

Spectroscopic analysis of complex objects is impossible without using the chemometric methods of analyzing experimental data. One of the main directions of chemometrics is using the mathematical, statistical, and other numerical methods to obtain the most relevant chemical information by analyzing spectroscopic data (Malinowski, 1980). Further, we will discuss the analysis of multicomponent spectroscopy data to demonstrate methods for determining the number of components.

Mathematical chemometric algorithms, actively developed after 1970 have been reviewed in the publications, the number of which exceeds dozens, or even hundreds of thousands. Multidimensional empiric relationships and structural features of the object.

In this field, the point is the development and application of mathematical algorithms designed to separate spectrum sets of multicomponent mixtures into items of individual compounds. At the same time, the well-known chromatographic and mass spectrometric methods are powerful, but they are not good enough to study equilibrium systems.

The field of the separation method of additive spectra data is large enough. By additive data can be meant the simultaneous measurement data of different characteristics of objects with the help of measuring instruments or multi-channel sensors, proportionally registering physical or any other parameters. The term "spectrum" will represent a set of such measurements in the form of dependence of "result of measurement – the ordinal number of the sensor channel".

There are different parameters, which lead to changes in the spectra of the mixture components. In chemistry, the parameters are the following: a time and conditions of chromatography, *i.e.,* pH of the medium, temperature, concentration of initial substances, etc. when studying equilibrium chemical systems; conditions of vital activity of biological objects; time and other parameters when researching intermediate products of multistage reactions; excitation energy in luminescence analysis and photoelectronic experiments; ionic strength of a solution and many others.

Separation methods of the multicomponent spectra (MCS) into the individual component spectra can be divided into two groups. One group allows one to find a unique set of spectrum components that coincides with the reference true ones. As to the other methods, using some *a priori criteria*, it is possible to find approximate contours of the individual spectra from the MCS. The application of the different techniques of the MLS expansion into individual spectra is determined by the nature of the information used, which, in turn, can also be divided into two groups.

Further, we will refer to the spectroscopic data matrix as the information source. Any other information is the additional one, which may concern either the relative component concentrations in the mixtures, the shape of the individual spectra, or the presence of known components.

Even in the simple case of the single mixture spectrum, its decomposition over the known components requires using the solution quality criteria, which would not be confined only by calculation of the Chi 2-value, but contain more details of the decomposition result reliability. In practice, the spectroscopic characteristics of the individual components are often unknown. In this case, the MCS can be decomposed into several individual spectra. New possibilities open up in opposition to a sole MCS, a set of several MCS with different component compositions has to be used. In addition to the non-negativity of the individual spectra, the main requirement is that the individual spectra additivity holds. For example, in the case of IR spectra of organic compounds, the molecules of the components should not form chemical bonds between themselves, which would lead to a shift in the frequencies of the spectral bands and a change in their shape (Hyvarinen, 2001). As to additional information that concerns the mathematical operational processing of the additive signal decomposition, one can refer to the work (Lee, 1998).

1.2. Theoretical fundamentals of the multicomponent spectra decomposition

Mathematical methods of decomposition of the additive spectral contour sets are based on the construction of linear combinations, *i.e.,* addition and subtraction with some coefficients of the observed MCS.

Let the contours of the spectra be represented on some grid of N values of the abscissa $i = 1, ..., N$ (in the future, we will not be interested in the absolute values and units of measurement of the abscissa scale, the data are assumed to be dimensionless). We denote the vector-spectra of mixtures $\boldsymbol{d}_j$, $j = 1, 2, ..., M$ (M is the number of mixtures in the studied multivariate set), the spectra of components are $\boldsymbol{e}_k$, $k = 1, 2, ..., K$ (K is the number of components). The additivity condition means that each of the mixture spectra must be a linear combination of the component spectra (for simplicity, we will not consider the influence of noise here):

$$d_{ij} = \sum_{k=1}^{K} e_{ik} \cdot c_{kj}, \qquad (1.1)$$

where e_{ik} is the relative spectral contribution of the kth component to the jth spectrum of the mixture with the weight (concentration) c_{kj}. The index i denotes the serial number of the point in the spectrum. In matrix form, the linearity relation is written as

$$D = E \cdot C, \qquad (1.2)$$

where D is the N x M matrix of mixture spectra, E is the N by K matrix of component spectra, and C is the K by M matrix of relative concentrations.

The most common approach is self-modeling, first proposed by Lawton (1971) to decompose two-component mixtures. Self-modeling is based on the representation of the component spectra x_{ik} as a linear combination of the eigenvectors $\boldsymbol{u}$ of the matrix of second moments (covariance matrix)

$$D \cdot D^T = U \cdot \Lambda \cdot U^T, \qquad (1.3)$$

Λ is a diagonal matrix of eigenvalues ordered in descending order:

$$x_{ik} = \sum_{j=1}^{M} u_{ij} \cdot y_{jk} \qquad (1.4)$$

where y_{jk} are the unknown coefficients, the index i refers to the number of the spectrum point, k is the component index, and j is the eigenvector index.

The vector representation of spectroscopic information taking into account relations (1.1) and (1.4) makes it possible to demonstrate the peculiarities of the problem. If the number of spectral points N exceeds the number of components K, a set of the component spectra vectors is defined in the N-dimensional space, K – dimensional subspace, which contains all spectra of mixtures (see (1.1)). *Figure 1.1* shows the case of $N = 3$, $M = 4$, $K = 2$, i.e., the subspace of the mixture spectra represents a plane in 3-dimensional space (intensities are plotted along the coordinate axes), and the component spectra vectors form the basis of this plane (non-orthogonal in the general case).

When matrix D is only known, the problem of determination of the component spectra matrix E reduces to finding the unknown matrices E and C involved in (1.2) in the form of a system of nonlinear equations, the solution of which is not unique. Indeed, if X' and C' are solutions of (1.2), it is always possible to find a nondegenerate matrix A, so that

$$D = X'' \cdot C'', \qquad (1.5)$$

where $X'' = X' \cdot A$ and $C'' = A^{-1} \cdot C'$. This problem of non-uniqueness is the main one in analyzing multicomponent mixtures by their spectra and is solved by involving additional information and restrictions on the type of solution.

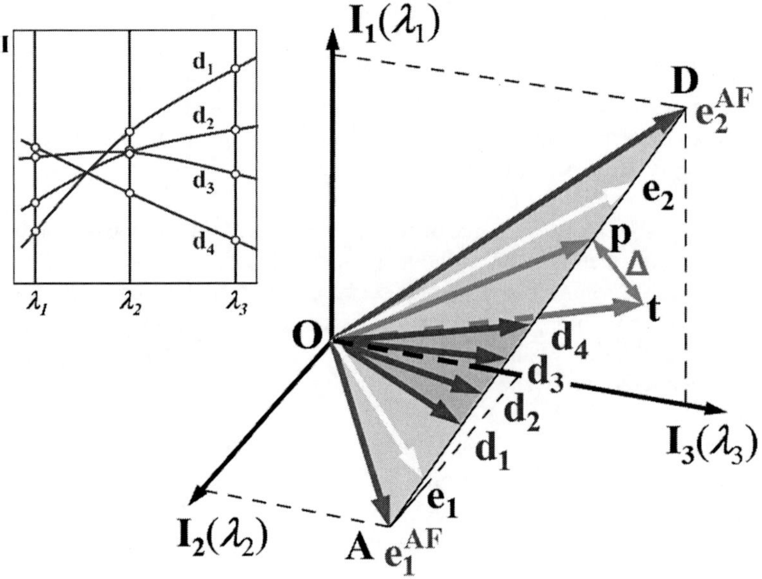

Figure 1.1. Vector representation of the decomposition problem for the case of 2-component spectra of 4 mixtures recorded at three wavelengths $\lambda_1 \div \lambda_3$. $\boldsymbol{d}_1 \div \boldsymbol{d}_4$ are the mixture spectra. $\boldsymbol{e}_1$ and $\boldsymbol{e}_2$ are the spectra of components. $\boldsymbol{e}_1^{AF}$ and $\boldsymbol{e}_2^{AF}$ are the non-overlapped component spectra. Along the axes, $I_1 \div I_3$, the spectrum intensities are deposited at three values of the abscissa $\lambda_1 \div \lambda_3$. $\boldsymbol{t}$ is a vector that is not a component of the mixture, $\boldsymbol{p}$ is its projection on the plane of mixtures AOD, Δ is the difference $\boldsymbol{t} - \boldsymbol{p}$. All vectors, except $\boldsymbol{t}$, are normalized for clarity so that their ends lie on the line AD. The areas of acceptable solutions are represented by sectors AO-$\boldsymbol{d}_1$ and DO-$\boldsymbol{d}_4$.

As is known, the common restrictions are the non-negativity of the spectra of components and their concentrations (Lawton, 1971; Sawal, 2014). Since the mixture spectra are non-negative linear combinations of the component spectra, regions of the solution existence are limited by the subspace regions defined by the mixture spectra vectors (2D-plane AOD in *Figure 1.1*). Accordingly, they are located in the 1st orthant inside a cone formed by the mixture spectra vectors (regions AO-d_1 and DO-d_4 in *Figure 1.1*).

It is shown in (Fok, 1973) that when the conditions of non-overlapping (called the Alentsev-Fok conditions) are met, the vectors of linear combinations that are most distant from each other correspond to the actual component spectra.

Consider some practical approaches for the decomposition problem.

Several methods of self-modeling curve resolution, such as RADICAL, SIMPLISMA (Rodionova, 2005), MCR-ALS (Mas, 2011), have become traditional. Recently, algorithms implementing the MILCA (Mutual Information Least Component Analysis) and SNICA (Stochastic Non-Negative Independent Component Analysis (Astakhov, 2006)) methods have been proposed. To date, the MILCA and SNICA algorithms have been successfully validated for multicomponent analysis of complex objects. It should be noted that the software packages implementing the MILCA and SNICA algorithms are available to users. Detailed Internet resources, including tutorials (URL http://www.ihes.fr/~zinovyev/#books) have been prepared on issues related to multivariate data analysis. In (Ohta, 1973), the self-modeling technique is modified for processing three-component mixtures. The authors (Meister, 1984) note that for four or more components, the algorithm is significantly complicated and has not been developed by them.

The non-negativity conditions are fundamental in the factor analysis, which is widely used under the decomposition of the mixture spectra (Malinowski, 1980; Martens, 1979). The latter is based on the possible linear combinations of eigenvectors. In (Martens, 1979), the main provisions of the factor analysis are formulated.

In several works, additional requirements to the shape of individual spectra are used to obtain a single solution, which ensures the successful operation of modern software packages, such as the already mentioned MILCA and SNICA. Thus, the authors (Kawata, 1985; Friedrich, 1987) suggest supplementing self-modeling with the requirement of the minimum information entropy of the component spectra, which is calculated under the assumption that the absolute value of the first derivative of the spectral contour p_{ik} means the probability density function, and then the entropy of the kth spectrum is equal to

$$H_k = -\sum_{i=1}^{N} p_{ik} \cdot \log_2 p_{ik} \cdot \qquad (1.6)$$

Authors (Friedrich & Yu, 1987) also use other complexity criteria. These are either the value of

$$\int_{\nu}\left[1+\left(\frac{\partial x(\nu)}{\partial \nu}\right)^2\right]^{1/2}, \quad (1.7)$$

where $\boldsymbol{x}(\nu)$ is the spectral contour.

The method provides the weakly overlapped areas in the spectra via similar ones in the contours of the eigenvectors U and the linear combinations of the latter to obtain the pure zero or resulting low-intensity spectra. This method is based on the Alentsev-Fok procedure (Fok, 1973) but applied to the eigenvectors U of the covariance matrix (1.3). The degree of the approach solution correctness of the component spectra depends on their non-overlap degree. The technique has been successfully applied to the 2- and 3-component mixture spectra decomposition.

The known decomposition methods turn out to be insufficient when the non-overlap zones are too short, or their number is not enough following to Alentsev-Fok (Fok, 1973).

The conditions of non-negativity of spectra and concentrations of components are successfully applied when separating small-angle X-ray scattering data in the **REGALS** package (Meisburger, 2021). The scattering intensity data is obtained, in particular, in the process of preparative chromatography of the sample in the stop-flow mode and then processed by a method called Evolving Factor Analysis (EFA) (Keller, 1992). From a set of experimental scattering intensity curves, sections with different numbers of components are selected that are estimated by analyzing graphs of singular numbers of data matrices, the size of which is increased by sequentially adding new measurements. After that, the individual scattering spectra are calculated using a variant of the ALS (Alternating Least Squares) method of non-negative least squares.

Interactive algorithms to analyze the spectra of mixtures are very effective in the case of a small (3-4) number of components (Bu, 2000; Leger, 2002).

There are several publications in which geometric analysis, like *Figure 1.1*, extended to three components, is effectively used to separate spectra (Akbari, 2013; Rajko, 2005, 2006). In what follows, we will consider approaches that are not constrained by the number of components.

1.3. Determination of the number of components

To determine the number of components K from the spectroscopic data matrix D can be regarded as a preliminary analysis. The known methods, as a rule, are based on knowledge of the magnitude of experimental errors and are essentially statistical (Lawley, 1971). In addition, several non-statistical methods have been proposed (Culler, 1984; Knorr, 1981; Shrager, 1982), which allow us to obtain plausible results. As to a real problem, it seems necessary to use several criteria since sole statistical ones may be incorrect due to the erroneous statistical hypotheses concerning experimental errors, and non-statistical estimates usually do not have sufficient justification. Most of the well-known methods of estimating K are based on a factor analysis of the matrix of initial data, which allows us to identify a set of "significant factors", the number of which is the number of components. Thus, in (Malinowski, 1980), more emphasis is placed on the practical use of factor analysis, and in (Lawley, 1971) – on the statistical aspect of the methods.

The estimation of the number of components can be done using the singular decomposition of the data matrix D:

$$D = U \cdot \Lambda \cdot V^T , \tag{1.8}$$

where U is an N x M matrix of eigenvectors of $D \cdot D^T$ (in the literature devoted to the factor analysis, it is a covariance matrix). V is the eigenvector matrix of the correlation matrix $D^T \cdot D$, and Λ is a diagonal matrix of singular numbers arranged in the non-increasing order (positive square roots from non-zero eigenvalues of the matrix or). Afterward, we will assume that the number of spectral points N, the number of mixtures M, and components K satisfy the inequality

$$N > M > K , \tag{1.9}$$

although the inequalities are sufficient

$$N \geq K \text{ and } M \geq N , \tag{1.10}$$

To facilitate the systematization of methods for estimating the number K, let us review the decomposition methods of (1.8).

Following to (Malinowski, 1980), we will present the experimental data matrix D in the form of the sum of the unperturbed matrix $D^*$and the matrix of experimental errors F:

$$D = D^* + F . \tag{1.11}$$

Under the condition (1.9), the rank of $D^*$is equal to K (in other words, the rank of the matrix D^* coincides with the rank of the relative concentration matrix C, this condition is assumed to be in further). But, due to the uncorrelated experimental errors, the rank of D is equal to the number of mixtures M.

In turn, the error matrix can be represented as

$$F = F^{\#} + F^0 , \tag{1.12}$$

where $F^{\#}$ is formed by the part of errors that contribute to the first K singular vectors U, and F^0 is composed of errors that are responsible for non-zero singular numbers λ_i, $i = M+1, \ldots, M$ and the corresponding vector columns u_i. Directly from (1.11) and (1.12), one obtains:

$$\begin{aligned} D^{\#} &= D^{*} + F^{\#} \\ D &= D^{\#} + F^{0} \end{aligned} \tag{1.13}$$

The matrix F^0 is usually called the residual error, which can be discarded by reconstructing the data matrix $D^{\#}$ from the first singular vectors of the original matrix D according to the relation (1.2). It should be noted that the initial errors are distributed over the matrices $D^{\#}$ and F^0, *i.e.*, they are formed only by some part of the errors.

The first group of methods for finding the number of components may be reduced to finding the splitting (1.13) (these methods are called residue analysis). At the first stage, the decomposition (1.8) is calculated, and then the singular vectors are divided into two groups, called primary and secondary axes. The primary axes are the first K singular vectors that form the basis of the column space (in the case of $\boldsymbol{u}$ – vectors) or rows (in the case of $\boldsymbol{v}$-vectors) of the matrix $D^{\#}$. If $Q = U \cdot S$, then

$$D = Q \cdot V^T = Q^{\#} \cdot V^{\#T} + Q^0 \cdot V^{0T}, \tag{1.14}$$

where $Q^{\#}$ is a matrix of primary axes of size N by M, whose columns $\{K+1, \ldots, M\}$ are zero, and Q^0 is a matrix of N by M secondary axes with zero columns $1,2, \ldots, K$. Then, the first method for estimating the number of components (Residual Standard Deviation, RSD) will consist in the calculation of the expressions

$$\text{RSD} = \sum_{i=1}^{N} \sum_{k=K+1}^{M} (q_{ik}^0)^2, \tag{1.15}$$

where q_{ik}^0 are the elements of the matrix Q^0;

$$\text{RSD} = \left(\frac{\sum_{k=K+1}^{M} \lambda_k^2}{N \cdot (M - K)} \right)^{1/2}, \tag{1.16}$$

(λ_k are singular numbers), which is equivalent to (1.9). Calculations are performed by changing the hypothesis about the number of components K starting from $K = 1$ and comparing the corresponding RSD value with the value of the experimental error estimate σ_{exp}. The value of K that leads to the RSD $\approx \sigma_{exp}$ is the number of components in the mixtures.

In practice, it is better to overestimate the number of components continuing the evaluation procedure until the RSD becomes strictly less than the experimental error. In (Bulmer, 1973), a method similar to the considered above is proposed, called the Residual Standard Deviation Method, based on the calculation

$$\sigma_K = \left(\frac{\operatorname{trace}\left(D^T \cdot D\right) - \sum_{k=1}^{K} \lambda_k^2}{N - K} \right)^{1/2}, \tag{1.17}$$

where K also takes increasing numbers until σ_K becomes compared with the value σ_{exp}. The main difficulty is caused by the correct estimation of the experimental errors that are not due to random noise with known dispersion, but may be systematic errors (instrument drift, intermolecular interactions, etc.). Moreover, the estimates considered are developed for the case of normally distributed noise.

It can be shown that the RSD estimate is related to the total quadratic discrepancy between the original D and $D^{\#}$ data matrices reconstructed by (1.13):

$$\mathrm{RSD} = \left(\frac{\sum_{i=1}^{N} \sum_{j=1}^{M} \left(d_{ij} - d_{ij}^{\#}\right)^2}{N \cdot (M - K)} \right)^{1/2}. \tag{1.18}$$

The RSD value is applied in the frequently used Factor Indicator Function (IND) (Culler, 1984; Knorr, 1981):

$$\mathrm{IND} = \frac{\mathrm{RSD}}{(M - K)^2}, \tag{1.19}$$

which is calculated by varying the estimate of K from 1 to M. The curve of the dependence of the IND value on M should reach a minimum value when estimating the number K is correct.

The natural method of estimating K seems to be based on comparing the variance of the elements of the residual matrix with the variance of the experimental errors. This method is implemented in some software packages. In practice, it can lead to incorrect results since the variance equality criterion of the two sets of random variables using the Fisher-Snedekor distribution is very sensitive to violations of the distribution normality (Johnson, 1977). In turn, as noted above, there is no reason to consider the distribution of errors in the data matrix normal. For the *F* criterion to be applied, it is first necessary to prove the hypothesis about the normality of the distribution law using the consent statistical criterion (Pearson's, Kolmogorov's, Smirnov's ones, see, *e.g.,* Johnson & Leone, 1977). Therefore, it is better to use statistical methods free from the distribution law knowledge of the error matrix elements or stable to deviations from distribution normality.

Here it is necessary to make an important remark. Quite widely used statistical analysis for the absence of correlations in elements of the residual matrix F^0, which corresponds to the difference between initial data matrix and matrix reconstructed by first singular vectors with indices $j=1, ..., K$:

$$F^0 = D - D^0, \qquad D^0 = U_{j=1,..K} \cdot \Lambda \cdot V^T_{j=1,..K}, \qquad (1.20)$$

which theoretically should contain only the noise component. In practice, it can give an underestimated number of components since the low-intensity component spectra are masked by noise and "captured" in F^0. Due to this circumstance, one will not discuss this method further.

A requirement for the residuals matrix is the absence of autocorrelation on the sequence of elements of its columns. The Durbin-Watson criterion (Durbin, 1971) can be used as the statistical test for residuals, which is governed by the formula:

$$\mathrm{DW} = \frac{\sum_{i=2}^{N \cdot M} \left(F_i^0 - F_{i-1}^0\right)^2}{\sum_{i=1}^{N \cdot M} \left(F_i^0\right)^2}, \qquad (1.21)$$

where F_i^0 represents a consecutive element of a matrix transformed into a one-dimensional array of size *M* by *N* by consecutive connection of vector-columns: $i = (1,1); (1,2); ... (1,M); (2,1); (2,2); ... (2,M); ... (N,M)$.

The criterion value lies in the region of {0.0 – 4.0}, taking values of about 2.0 in the case of absence of correlation. If the hypothesis of the existence of a correlation between consecutive elements of the array has a significance level of 0.05, it should be rejected if the value of the criterion lies in the range {1.7 - 2.2}.

In practice, it is necessary to assume the presence of a weak residual correlation due to a violation of the additivity of the spectra during their interaction in mixtures or unaccounted weak systematic measurement errors. In most cases, the boundary needs to be slightly expanded, for example, to {1.5 – 2.5}, justifying this expansion empirically based on the experience of solving similar problems from this subject area.

Another criterion is based on calculating the variance of the eigenvalues of $D \cdot D^T$ or $D^T \cdot D$ matrices. According to the definition of the singular decomposition (1.8), these eigenvalues are the squares of the corresponding singular values.

The standard error in the *m*th eigenvalue is calculated as (Bulmer, 1973):

$$\sigma_m = \left(\sum_{j=1}^{M} \sum_{k=1}^{M} v_{mj}^2 \cdot v_{mk}^2 \cdot \sigma(Z)_{jk}^2 \right)^{1/2}, \tag{1.22}$$

where v_{mj}^2 и v_{mk}^2 are *j*th and *k*th elements of the *m*th singular vector, and

$$\sigma(Z)_{jk}^2 = \left(\begin{array}{ll} \sum_{i=1}^{N} \left(d_{ij}^2 \cdot \sigma_{ik}^2 + d_{ik}^2 \cdot \sigma_{ij}^2 \right) & j \neq k \\ \sum_{i=1}^{N} \left(4 d_{ij}^2 \cdot \sigma_{ij}^2 \right) & j = k \end{array} \right)^{1/2}. \tag{1.23}$$

The value σ_{ij} denotes the error in the experimental point d_{ij}. The number of components is the number of eigenvalues exceeding the estimate (1.22) at a given significance level. As seen from (1.23), this method allows us to take into account the error estimate in each of the spectral points of the data matrix *D* individually.

When the standard deviation varies from point to point in the spectrum and is not a constant for a given data matrix, one suggests applying the chi-square criterion (Bulmer, 1973). The disadvantage of this method is the necessity to have a fair error estimate at each experimental point. As are obtained, they are defined as

$$\chi_N^2 = \sum_{i=1}^{N} \sum_{j=1}^{M} \frac{\left(d_{ij} - d_{ij}^{\#} \right)^2}{\sigma_{ij}^2} \tag{1.24}$$

where $d_{ij}^{\#}$ are still the elements of the data matrix reconstructed from the first *K* singular vectors (1.2). The number of components *K* is determined by changing it, starting from 1, calculating χ_N^2 for each trial value *K*, and comparing it with the expected value χ_α^2

$$\chi_\alpha^2 = (N-K)(M-K), \tag{1.25}$$

where $(N-K)(M-K)$ is the number of degrees of freedom, α is the significance level. Choose such a value of *K*, which χ_N^2 is closest to χ_α^2 (1.25).

From a practical viewpoint, methods based on the spectrum of singular numbers λ_j of the matrix D analysis (or the eigenvalues of the covariance matrix (Knorr, 1981)) are of great interest. As is shown in (Ohta, 1973; Culler, 1984; Gillette, 1982), due to the normal distribution of errors, the singular numbers of the error matrix F from (1.11) lie about on a straight, when they are arranged in descending order on the graph in logarithmic magnitudes against ordinal numbers. The mean square deviation from this line proposed to be estimated by the formula (Ohta, 1973):

$$\xi_0 = 0.3 \cdot \sigma_D^2 \cdot \sqrt{N \cdot (M-2)}, \tag{1.26}$$

where σ_D^2 is the estimate of the experimental error variance. The deviation of the eigenvalues from the straight line on the graph becomes the more significant, the greater the dependence of σ on d_{ij} (for example, it may be in the case of Poisson noise), which sometimes limits the application of this method. The number of singular values falling out of the linear dependence is the number of components. This semi-empirical method is very convenient. In the REGALS package (Meisburger, 2021), for example, the index of the smallest singular number λ_j is taken as the number of components for which the condition is still met

$$\frac{\lambda_j - \sqrt{M}}{\sqrt{N}} > 1.0, \tag{1.27}$$

where M is the number of points in the spectrum, N is the number of the mixture spectra. However, this approach is valid for data in which the variance expectation does not depend on the abscissa, so the resulting estimates should be treated with caution.

Among the methods for estimating the number of components that do not require knowledge of the experimental error value, we should note the frequently used Embedded Error Function (IE) (Culler, 1984):

$$\mathrm{IE} = \left(\frac{K \cdot \sum_{j=K+1}^{M} \lambda_j^0}{N \cdot M \cdot (M-K)} \right)^{1/2}, \tag{1.28}$$

which is calculated for the numbers K changed from 1 to M. The curve of the IE-dependence on K looks like the log (λ_j)-dependence on the number j. The number of components corresponds to the breaking point on the curve that corresponds to the sharp change in the slope angle of the curve relative to the abscissa axis).

The number of components can be found by a frequency analysis of the contours of the left singular vectors. As is noted in (Shrager, 1982), singular vectors corresponding to "significant" singular values contain mainly lower harmonics (in the Fourier spectrum). The "noise" vectors represent the high noise harmonics. The latter is valid only if the noise components have a higher frequency than the spectroscopic information data. Therefore, under the number of components estimated by this method, it is essential that the spectra sampling is carried out with a sufficiently small argument step (wavelength). In turn, the sampling noise will exceed its upper Fourier spectrum frequency.

The two other criteria for checking individual spectra included in the mixture spectra, namely, formulated in (Malinowski, 1981), are Reliability (RELI) and Spoil (SPOIL) ones. The criteria, developed for the library search of the spectrum components, show reliable results when estimating the number of components by projecting the library-known spectrum onto the data matrix $D^{\#}$ column space. For this purpose, the matrix $D^{\#}$ is reconstructed from the *K*-first singular vectors of the matrix *D* (1.8). These criteria use the "apparent" error value in the vector $\boldsymbol{t}$ of the trial spectrum of the component:

$$\mathrm{AET} = \left(\frac{\sum_{i=1}^{N} (p_i - t_i)^2}{N} \right)^{1/2}, \tag{1.29}$$

where $\boldsymbol{p}$ is the vector of the projection of $\boldsymbol{t}$ on *D* (or $D^{\#}$) (see *Figure 1.1*), which may be calculated using the SVD decomposition (1.2):

$$\boldsymbol{p} = U \cdot U^T \boldsymbol{t}, \quad \text{or} \quad \boldsymbol{p} = U^{\#} \cdot U^{\#T} \boldsymbol{t}, \tag{1.30}$$

where the sign # denotes the matrix composed of the first *K* columns of *U*, and *K* is the assumed (verifiable) number of components.

The value of AET is compared with the "real" error in the projection vector $\boldsymbol{p}$ (λ_j are the singular numbers from (1.2)):

$$\mathrm{REP} = \left(\frac{\sum_{j=1}^{M} \lambda_j^2}{\max(N, M) \cdot (M - K)} \right)^{1/2}, \tag{1.31}$$

$$\mathrm{RET} = \left(\mathrm{AET}^2 - \mathrm{REP}^2 \right)^{1/2}.$$

The criteria used in deciding whether a test vector is a component are calculated using the formulae:

$$SPOIL = \frac{\text{RET}}{\text{REP}}, \qquad \text{RELI} = \left(\frac{\text{RET}^2 - \text{RET}_{\text{est}}^2}{\text{AET}^2} \right)^{1/2}, \tag{1.32}$$

where the RET_{est}^2 is the estimated error of the data used for building the test vector.

According to (Lawton, 1971), the following threshold values are proposed: RELI > 0.5 means that ***t*** (or a set of the test spectra) is acceptable as the spectrum of the component, as well as SPOIL < 3.0. SPOIL = 3.0...6.0 indicates the possibility that ***t*** is the component spectrum; SPOIL > 6.0 means a negative answer.

Despite the apparent rigidity of the error theory, based on the "triangle ratio":

$$\text{AET} = \text{RET}^2 - \text{REP}^2.$$

This equality is introduced artificially and not supported by any statistical conclusions. Notice we must consider the RET and REP errors as they would be independent, and their variances could be summed. Nevertheless, the considered criteria allow one to obtain good results.

Thus, there is a wide variety of methods for estimating the number of components, the simultaneous use of which makes it possible to increase the reliability of the answer by comparing the results obtained by different methods. The very fact that the estimates do not match can carry useful information. In opposition to the statistical techniques, the overestimated number of components obtained by the non-statistical ones can evidence that errors caused, *e.g.*, by the temporary instability of the device (wavelength drift, *etc.*) are predominant. In turn, intermolecular interactions in mixtures lead to changes in the spectra of components, *etc.*

In the next Section, several examples of estimating the number of components are presented.

1.3.1. Data preparation

Several mixtures of organic compounds were prepared to demonstrate the procedures for determining the number of components. Vacuum distillation and preparative liquid chromatography were used. The chromatography device was a 250x3.5 mm glass column filled with Silasorb 600 (LC) sorbent of 30 microns, equipped with a syringe sample inlet and pressure system (a nitrogen cylinder with a pressure reducer for 0-0.4 MPa overpressure). The flow rate was 0.75 ml/min. The device design is simple, and we do not give its scheme since it is not of fundamental importance. Perkin Elmer 580B IR spectrometer coupled to an Interdata 6/16 microcomputer was used as a detector.

The spectra of the mixtures-fractions were taken in a flowing cuvette of KBr with a thickness of 0.15 mm, stopping the solvent flow with a needle valve at approximately equal intervals. The level of spectroscopic noise was 0.4% T. The obtained spectra have been smoothed using standard programs (included in the spectrometer software). The spectrum of the solvent (carbon tetrachloride, which was a known component) was compensated using a cuvette of variable thickness placed in the reference spectrometer channel or by a program for spectra subtracting. The initial data was formed by encoding the most informative sections in the mixture spectra with a variable step. The relative component concentration matrix has been determined onto the chromatographic peaks obtained on a gas chromatograph LXM 8MD, namely: the carrier is 3%E-30 on Panchrome 3, column length 3m, temperature 120oC, with an accuracy up 3.5% rel. (Volkov, 1996).

The model mixtures of hexane, toluene, acetone, cyclohexanone, diethyl ether, and amyl acetate (a model of technical waste of polymer production) have been used. Chromatograms have been recorded at the frequency of 1450 cm^{-1} when absorption is approximately the same for all the components. The example of a typical chromatogram is shown in *Figure 1.2*. The spectra of the mixtures have been recorded in the range of 1750-1110 cm^{-1} with a resolution not worse than 5 cm^{-1}. After that, when preparing the spectroscopic data matrix D, the low-intensity areas were discarded, and the regions 1750-1690, 1610-1595, 1510-1300, and 1160-1110 cm^{-1} were left.

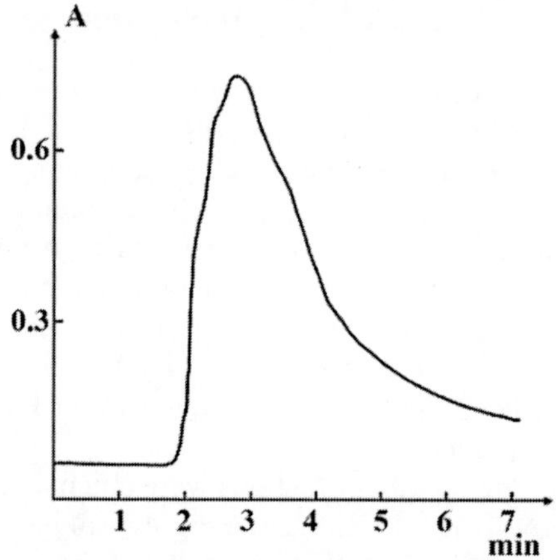

Figure 1.2. Typical chromatogram of a 5-component mixture **No. 1** of hexane, toluene, acetone, cyclohexanone, and diethyl ether at an absorption frequency of 1450 cm^{-1}. The monotony of the peak points to poor component separation under these chromatographic conditions.

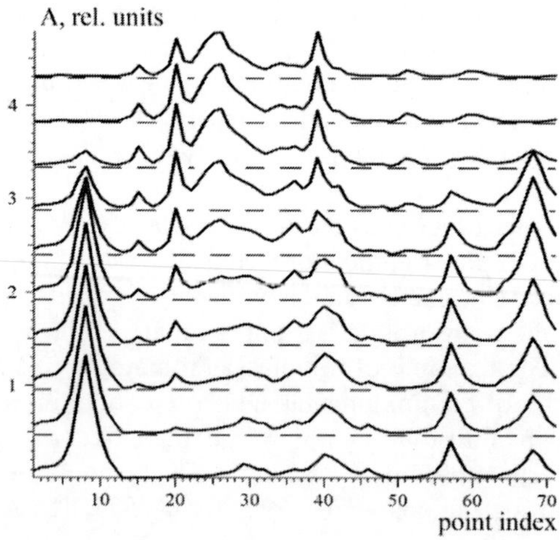

Figure 1.3. IR spectra of the 5-component mixture №1 after digitization and resampling. The spectral range is 1750-1110 cm^{-1}. The serial numbers of spectral points on the abscissa axis are given (since the wavelength value is not considered in the decomposition, the spectra are digitized with a variable step to exclude non-informative regions). The spectra are shifted vertically for better visualization, and the dashed lines indicate zero absorption levels.

The decoded spectra have been digitized with the fixed step of 5 cm^{-1}. *Figure 1.3* shows the typical mixture spectra **No. 1**, the five components: hexane, toluene, acetone, cyclohexanone, and diethyl ether. The 10-12 temporal spectra of fractions have been selected from each mixture. To verify the decomposition performance is desirable to get all the precise parameters asap. Based on equation (1.1), the concentration matrix *C* was calculated using the pure component spectra measured on the same spectrometer.

The second (**No. 2**) model was a mixture of the six (6) components: isooctane, toluene, benzene, diethyl ether, acetone, and amyl acetate in carbon tetrachloride as a nonpolar solvent. In this set, two pairs of components are relatively strongly overlapped: acetone – amyl acetate and benzene – toluene (see *Figure 1.13*). In addition, these substances are poorly separated under these chromatographic conditions. The set of absorption spectra is shown in *Figure 1.4*.

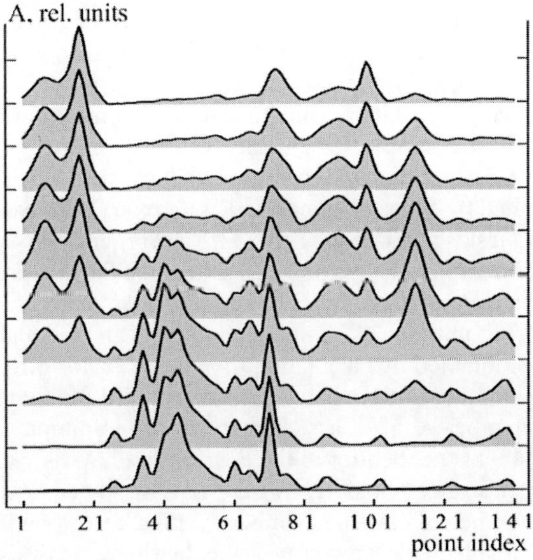

Figure 1.4. The IR spectra of fractions of the 6-component mixture (**№ 2**) of isooctane, toluene, benzene, diethyl ether, acetone, and amyl acetate. For other details, see the capture in *Figure 1.3*.

Model **No. 3** was composed of six (6) components: hexane, isooctane, benzene, dioxane, diethyl ether, and acetone. The mixture fractions (10 fractions) were obtained by distillation with a deflagrator. Determination of the number of components by matrix (*Figure 1.5d*) gave an estimate of 4 instead of 6.

Model **No. 4** was the 4-component mixture of acetone, diethyl ether, cyclohexanone, and toluene. The fractionation has been carried out using liquid chromatography. However, a worse component separation was obtained due to the larger column diameter (5 mm).

1.3.2. Analysis of singular values and residual matrices

As was described above, the preliminary estimate of the number of components is obtained by counting the number of singular numbers above the line drawn through the set of small $log(\lambda_k)$. By the break of the graph line in the transition from "noisy" $log(\lambda_k)$ to "significant" level, one can judge about nature of the spectroscopic information. Loosely speaking, one can perform the splitting into significant and noisy components more confidently since the decomposition procedure is better conditioned. *Figure 1.5* shows the singular values for several modeled and real data sets.

The second method, which can be efficient for estimating the number of components, is based on statistical criteria, which work not with the data values, but with their ranks, which make the estimates independent of the distribution law (Johnson, 1977). Rank criteria are invariant regarding any monotonic transformation of the measurement scale. One of the rank criteria was formulated by Mann-Whitney and Wilcoxon (Wilcoxon, 1945). Mann-Whitney's U test is a nonparametric statistical criterion to assess differences between two samples. The hypothesis stated that the distribution laws of the two independent random variable sets are identical can be used as the criterion of the number of components to test the residuals matrix $F_0(K)$. The latter is obtained by calculating the matrix $D^{\#}$ for different numbers K of the sample of the experimental error matrix (see equations (1.12)-(1.15)). As to the number K, it is assumed to be the minimum value when the hypothesis about the identity of the distribution laws is valid at the chosen level of significance. In other words, K is assumed to be equal to the minimum number of singular numbers, which corresponds to the residual matrix F_0 with elements having the same distribution law as the errors in D.

First, all n_1 elements of $F_0(K)$ and n_2 elements of the array of experimental errors are combined into one vector $\boldsymbol{x}$ in descending order. Then form a vector $\boldsymbol{y}$ from n_1 numbers representing order numbers of elements $F_0(K)$ in vector $\boldsymbol{x}$ and calculate the sum

$$S = \sum_{i=1}^{n_1} y_i \tag{1.33}$$

According to the tables (Johnson, 1977) for the given n1 and n2 and the selected significance level, we find the critical region (S_{min}, S_{max}). The value S belonging to the interval (S_{min}, S_{max}) means that the distribution laws are identical. For the IR spectra of 6-component mixtures shown in *Figure 1.4*, the S value for the 4-component hypothesis is 194, which is outside the interval (348, 472) at a 5% level of significance. For the 5-7 component hypotheses, the value S was equal to 429, 394, and 414, respectively. They may indicate the presence of errors associated with instrument drift and lead to the fact that the experimentally estimated noise vector contains a fraction of information about the spectra of the substances. Hence, the acceptable number of components is 5.

A slight modification of the Wilcoxon procedure makes it possible to create the criterion of equality of dispersions under the condition of equality of position characteristics for the two populations (Johnson, 1977). In our case, since one assumes the mean values of $F_0(K)$ and experimental errors to be zero, this criterion can be applied. The modification consists of changing the way of assigning ranks to elements of vector x. Instead of the sequence 1,2,3, ..., n_1+n_2, one constructs the following: 1,4,5,8,9, ..., 7,6,3,2 (this trick is to destroy autocorrelations in the sequence) and from this chain form the vector y of n1 ranks (ordinal numbers in x) for sample F0(K). Then, the number S is calculated, and the criterion is evaluated using statistical tables (Johnson, 1977). In the example under consideration, the values of S are 292, 362, and 440 for the 5, 6, and 7 component hypotheses. As is seen, the acceptable K is equal to 6, *i.e.*, the criterion of variance equality, in this case, was more effective.

It should be noted that it is better to apply both the above criteria together. The number K is chosen as a minimum estimation that is simultaneously valid for both approaches. Even the estimate of experimental errors is unknown, the number of components can be found from testing the hypothesis that supposes that the mean value of elements $F_0(K)$ is equal to zero.

The practice has shown that the analysis of the residuals matrix in the presence of autocorrelation gives less reliable results due to the influence of systematic measurement errors. The dependence of the autocorrelation criteria values on the estimated number of components under consideration has a significantly less pronounced kink than the plots of the logarithms of the singular numbers.

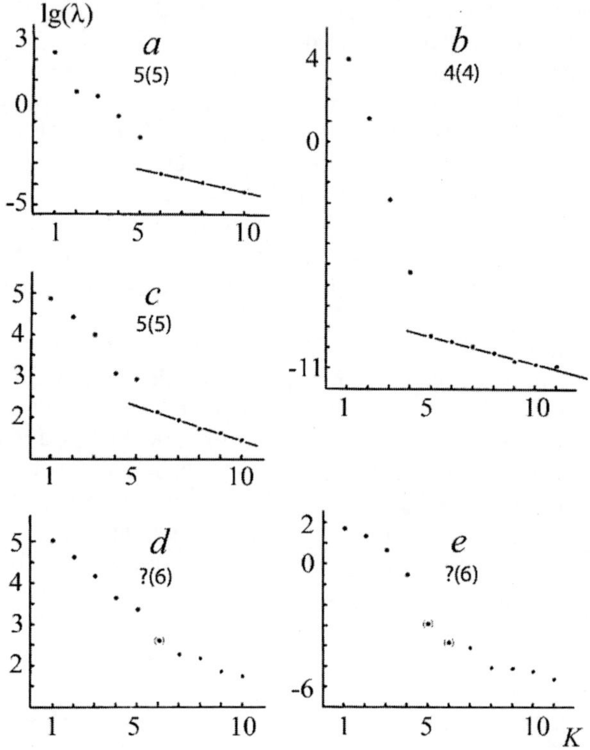

Figure 1.5. Logarithms of singular numbers of the IR spectra matrices of the test mixtures: a) the model data matrix at spectral noise level 0.2%T, b) mixture No.4, c) mixture No.1, d) mixture No.2, e) mixture No.3. Each plot shows the number of absorbing components found and the a priori known (in parentheses). The ordinal numbers of singular numbers are plotted on the abscissa. The question marks show that the nonlinear dependences occur at $R^2 > 0.7$, and the estimate is still a problem.

1.3.3. Analysis of singular vectors of the data matrix

The contributions of component spectra and spectroscopic noise to the singular vectors U of the data matrix are nonuniformly distributed over them. Let u_j be the jth column of the matrix U. u_1 represents an averaged mixture spectrum, which exhausts the maximum variance in the original matrix. By definition (1.8),

$$\boldsymbol{u}_j = \frac{1}{\lambda_j} D \cdot \boldsymbol{v}_j$$

$$\boldsymbol{v}_j = \frac{1}{\lambda_j} D^T \cdot \boldsymbol{u}_j \quad (1.34)$$

In addition, the DD^T and D^TD matrices are symmetric and non-negative (since $d_{ij} \geq 0$ – we consider the case of non-negative spectra). If there are no regions in the mixture spectra whose intensity is equal to zero for all the spectra, the DD^T and D^TD matrices are irreducible.

According to the Perron-Frobenius theorem (Lancaster, 1985), non-negative irreducible matrices have a positive eigenvalue λ_1

$$\lambda_1 = \sqrt{\max_{x \neq 0} \frac{|D \cdot D^T \cdot x|}{|x|}} = \sqrt{\max \frac{|D^T \cdot D \cdot y|}{|y|}}, \quad (1.35)$$

equal to the matrix spectral radius, and it refers to a positive eigenvector u_1 (and v_1, respectively).

As follows from (1.29), u_1 is the positive linear combination of the mixture spectra (matrix D columns), whereas v_1 is the positive linear combination of the matrix D rows.

If the matrix A_k of the smaller rank $k<M$ is obtained using the k first singular vectors

$$A_k = U \cdot \Lambda \cdot V^T, \quad (1.36)$$

the residual matrix $A - A_k$ has the minimal possible consistent norm for all $A - B_k$, where B_k is a matrix of rank k, i.e.

$$\|A - B_k\|_E = \min \quad \text{if} \quad B_k = A_k. \quad (1.37)$$

Consequently, when $k = 1$, the residual matrix after removing the 1-st singular vector will have the minimum possible element variance. It means that in u_1 or v_1 the signal-to-noise ratio will be maximum if the errors are uncorrelated. Similarly, u_2 and v_2 maximally exhaust the correlations in the residual matrix, and so on, until the vectors $\{u_K+1, ..., M\}$ and $\{v_K+1, ..., M\}$ are composed of elements whose non-zero values are due almost exclusively to the uncorrelated component in the raw data, in practice, to noise.

Figure 1.6 shows the left singular vectors calculated from a matrix of 10 mixture spectra (6-component mixture No.2, see 1.3.2) shown in *Figure 1.4*. It can be seen from this figure that the relative contribution of high-frequency components increases with increasing vector numbers. If the sampling interval in the data exceeds the maximum frequency in the spectra, the high-frequency part corresponds to the noise component.

An effective in-practice method for estimating the number of components is based on the statistical analysis of the sequence of elements in the left singular vectors. The condition for the applicability of this method is a sufficiently small step of spectra digitization. The latter ensures the correlation (monotonic change) of consecutive points in the non-noise spectra.

Consider now the analysis of autocorrelations in singular vectors, on the example of the set of 6-component IR mixture spectra (model **No. 2**, *Figure 1.4*). The left singular vectors are given in *Figure 1.6,* while *Figure 1.7* shows a plot of the logarithms of the singular numbers. As a result, one can see that the choice can be made between the 5 and 6 components. *Figure 1.6* is visually more informative: singular vectors 1 ÷ 6 describe the spectral data, while vectors 7 ÷ 10 have a noise origin.

Tables 1.1 and 1.2 show the statistical analysis results of the singular vectors using Wilcoxon and Durbin-Watson (1.21) tests. The results show that at the most commonly used significance level of 0.02, we should take the seven components instead of six ones as the number estimate.

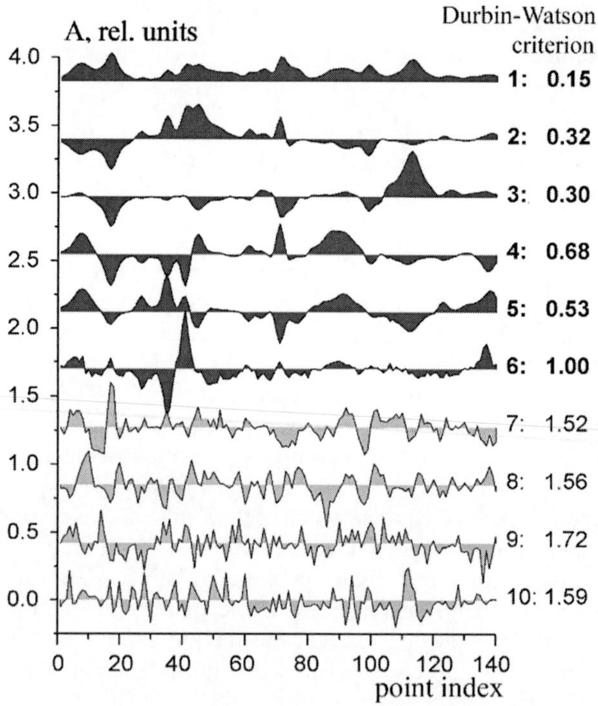

Figure 1.6. Left singular vectors u_j of the 6-component data matrix (*Figure 1.4*) presented as spectral contours. The contours are shifted vertically for better visualization, the zero-level lines are in the middle of the grey shading areas.

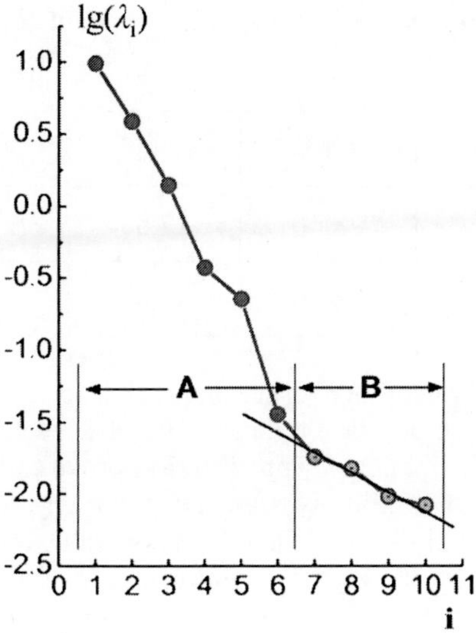

Figure 1.7. Plot of logarithms of singular numbers of the matrix of 6-component mixtures (model No. 2, see 1.3.2). The coefficient of determination R^2 for linear approximation of the right part of the graph is 0.89 > 0.7. Area A corresponds to significant numbers, B to the noise part.

Table 1.1. Values of the Wilcoxon and Durbin-Watson (1.21) criteria for autocorrelation of elements in the left singular vectors *U* calculated for the data set No.2.

Number of the singular vector	Wilcoxon test	The decision about the number of components (at significance level 0.05)	Durbin-Watson test (1.14)	The decision about the number of components (at significance level 0.05)
1	1.0	-	0.15	-
2	2.0	-	0.32	-
3	8.0	-	0.30	-
4	14.0	-	0.68	-
5	11.0	-	0.53	-
6	17.0	-	1.00	-
7	25.0	6 (0.002)	1.52	(6)
8	28.0	7 (0.02)	1.56	-
9	28.0	8 (0.02)	1.72	8
10	30.0	9 (0.05)	1.59	-

Table 1.2. Significance levels for the Wilcoxon test calculated for data set No.2.

Level of significance Q	0.05	0.02	0.01	0.002
Lower critical value	28.00	27.00	26.00	23.00
Upper critical value	43.00	45.00	46.00	48.00

The Durbin-Watson DW criterion (1.21) showed the overestimation of $K = 8$ due to weak correlations along the vectors due to molecular interactions of components. This criterion proved to be more sensitive to such violations of additivity. As stated earlier, the hypothesis of the presence of correlations is rejected if the DW value is outside the confidence interval {1.7-2.2} for the typical significance level $Q = 0.05$. However, if a confidence interval is enlarged to {1.5-2.4}, then the number 6 can be taken as an estimate of K.

The general conclusion for the example above is to estimate the number of components as 6 or 7. However, number 8 must also be considered when searching for individual spectra discussed later.

The Durbin-Watson criterion at $K = 8$ was rejected in this experiment because the contribution of the 8th singular vector in the data matrix can be neglected due to the relative smallness of the corresponding singular number (*Figure 1.7*).

In addition to the Durbin-Watson criterion, the nonparametric Wald-Wolfowitz series criterion (Wald, 1940) can be recommended. It allows one to obtain unbiased and consistent estimates for small sample sizes (parametric criteria usually require a size of 1000 at least) and *a priori* unknown distribution laws. The Wald-Wolfowitz statistic checks the randomness of alternation of positive (A) and negative (B) elements in the binary sequence AAABBBAAAAAABA (in our case, the signs of consecutive elements of singular vectors). Let N_A be the number of positive singular vector elements, N_B is the number of negative elements, and N_R is the number of groups of both.

The expectation and variance of the number of groups are equal to

$$M\{N_R\}=1+\frac{2\cdot N_A\cdot N_B}{N_A+N_B},\quad D\{N_R\}=\frac{2\cdot N_A\cdot N_B\cdot(2\cdot N_A\cdot N_B-N_A-N_B)}{(N_A+N_B)^2\cdot(N_A+N_B-1)}.$$

These values are used to test the null hypothesis that the sequence is not random.

The null hypothesis is rejected if inequalities take place

$$Z\left(1-\frac{Q}{2}\right)+\frac{1}{2\sqrt{D\{N_R\}}}\leq\frac{N_R-M\{N_R\}}{\sqrt{D\{N_R\}}}\leq Z\left(\frac{Q}{2}\right)-\frac{1}{2\sqrt{D\{N_R\}}}. \quad (1.38)$$

The criterion $\frac{N_R-M\{N_R\}}{\sqrt{D\{N_R\}}}$ has the normal distribution Z with zero expectation and unit variance. Q is the significance level. Thus, the Wald-Wolfowitz criterion is used to test the hypothesis that the consecutive elements of the left singular vectors are not a sample of a random sequence.

It is interesting that in physics experiments with bright radiation sources, such as 4th generation synchrotrons and free-electron lasers, sets of spectra, or scattering data, taken at specific time intervals, are obtained. It is possible to study the temporal change in the structure of samples, in particular, by mathematically extracting the components of spectral contours. In these experiments, there is also a severe problem of sample degradation under the influence of high-intensity radiation, and methods of component analysis can be applied in such cases in control programs, recognizing the moment of the beginning of degradation, which will be characterized by the growth of the number of components as new frames of measurements arrive

Examples of monitoring a set of frames during searching for sample stability conditions are shown in *Figures 1.8 ÷ 1.11* (small-angle scattering intensity data measured from a protein solution with the interval 15 sec.). The changes in the intensity curves can be controlled by using the criteria for one-dimensional shape difference and estimating the rank of the data matrix (composed of individual frames). Since the data frames must coincide, the rank of the measurement matrix must be unity. The singular analysis allows fixing the moment of the rank increase and providing more detailed information about the changes occurring in the sample. *Figure 1.8* shows that the changes in the shape of the scattering curves start almost immediately. The singular vectors in (*Figure 1.9*) indicate the presence of the two main components in the data set, as indicated by the hypothesis test criteria for the autocorrelation along the left singular vectors (Table 1.3).

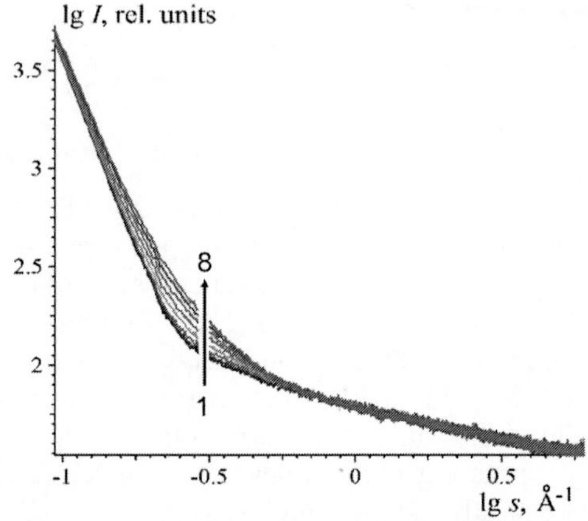

Figure 1.8. Data from 8 consecutive measurements of X-ray small-angle scattering from a sample of protein solution at synchrotron station X33 (EMBL c/o DESY, Hamburg, Germany). The scattering intensity in the central region increases with time. The enumeration of the frames goes from bottom to top. The values of the logarithms of the scattering vector modulus are plotted on the abscissa axis, $|\boldsymbol{s}| = \frac{4\pi \sin\theta}{\lambda}$, λ is the wavelength, and 2θ is the scattering angle.

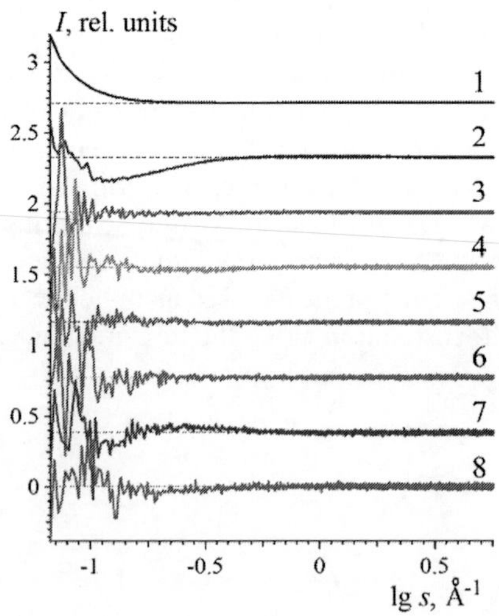

Figure 1.9. Left singular vectors u_j of the data matrix (*Figure 1.8*, presented as spectral contours. The contours are shifted vertically, and zero lines are shown separately for each one. The units on the axes are similar to *Figure 1.8*.

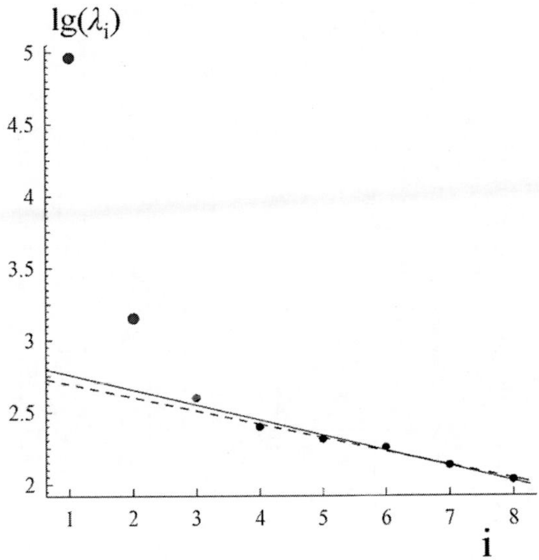

Figure 1.10. Logarithms of singular values of the data matrix (*Figure 1.8*). The R^2 criterion for linear approximation is of the order of 0.69 ~ 0.7 (solid line). Above the straight line, there are two singular numbers, which correspond to the number of significant components. However, more adequate in terms of the value of $R^2 = 0.92$ (dashed line), the estimate of the number of components is equal to 3.

Here it is necessary to give explanations. According to *Table 1.3*, choosing between 2, 3, 7, and 8 components is required. The general rule can be the choice of the minimum of the obtained estimates since $K = 7$ and 8 correspond to too small singular values (*Figure 1.10*). The Wald-Wolfowitz criterion for the nonrandom nature of elements of the 7th and 8th singular vectors is due to small systematic drifts of equipment and measurement conditions, so their contribution to the experimental data contours is small. Thus, we can conclude that the number of components in the data set equals two, and they probably correspond to dimeric and monomeric macromolecules in the studied solution.

Table 1.3. Values of the Wald-Wolfowitz (1.38) and Durbin-Watson (1.21) criteria calculated for singular vectors of the small-angle X-ray scattering data set (*Figure 1.8*).

Number of the singular vector	The value of the Wald-Wolfowitz criterion (1.32)	Lower and upper significan-ce levels for (1.32) at P=0.95	Conclu-sion on the number of compo-nents	The value of the Durbin-Watson criteri-on (1.14)	Conclu-sion on the number of compo-nents (at 0.05 significan-ce level)
1	1.0	0	-	$1.6*10^{-2}$	-
2	0.88	$9.4*10^{-3}$ $4.7*10^{-2}$	> 1	$7.1*10^{-2}$	> 1
3	-0.24	-1.98 1.98	**2**	1.4	> 2
4	-1.79	-1.98 1.98	< 4	2.1	**3**
5	-1.56	-1.98 1.98	< 5	2.4	< 5
6	0.71	-1.98 1.98	< 6	1.9	< 6
7	-2.9	-1.98 1.98	**7?**	1.7	< 7
8	-2.2	-1.98 1.98	**8?**	1.7	< 8

Another method for evaluating the number of components is based on the window calculation of the autocorrelation criterion for elements of singular vectors. The criterion is calculated inside the scan window (the window should include no fewer than the 20 data points), and the obtained value is attributed to the central point of the window. The endpoints are calculated using reduced windows. A trapezoidal contour with a base ratio of 1:2 may be proposed as a scanning function. The graphs of the criterion values clearly show areas with systematic deviations (DW value (1.21) less than 1.7) and which correspond to random noises ($1.7 < DW < 2.2$). The boundary values shown here are for the typical case of the number of points in the tested sequence within a window of 30 ÷ 100 points and the significance level of the correlation hypothesis of 0.05.

Figure 1.11 shows the plots of the window DW criterion applied to the left singular vectors for the above example of low-angle scattering data analysis. The width of the scanning window in this example was 10% (210 points) of the entire data range. It can be seen from the picture that the graph of statistics for the third singular vector is close to the range of 1.7 – 2.2, and this vector can be considered a random noise. Deviations in the plots with numbers 4 – 8 outside the range $1.7 < DW < 2.2$ indicate the presence of low-intensity components, which can be attributed to instrumental distortions.

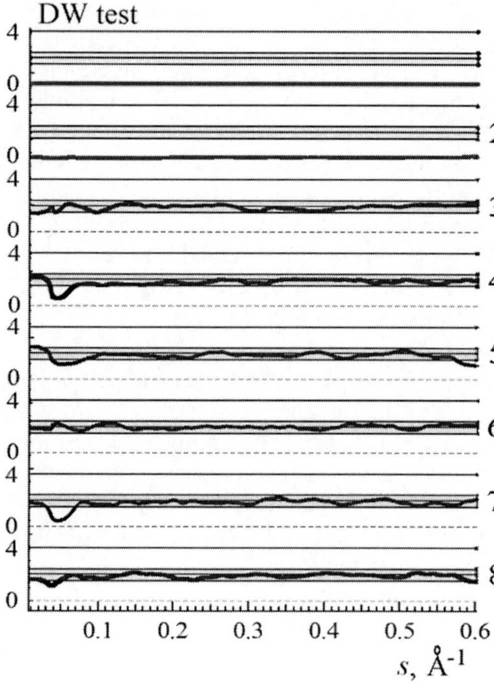

Figure 1.11. Graphs of the Durbin-Watson window statistics for the elements of the left singular vectors of the data matrix of *Figure 1.8*. Right-side numbers correspond to numbers of singular vectors. Each criterion graph is shown with a vertical offset in the frame corresponding to values $0 < DW < 4.0$, the levels of critical values 1.7, 2.0, and 2.2 (see text) are shown inside the frame by three horizontal lines marked by grey areas. The criterion values inside the range $1.7 \div 2.2$ correspond to the rejection of the hypothesis of autocorrelations in the scan window. The scan window length is 10% of the angular range shown on the abscissa axis.

1.3.4. The use of reference spectra

Another method for determining the number of components, which is quite convenient for expert work, is based on projecting the reference component spectrum present in the mixture into subspaces formed by an increasing number of left singular vectors of the data matrix. This operation is known in the literature on factor analysis as Target Transformation (Malinowski, 1980). Let $\boldsymbol{t}$ be the vector spectrum of a pure substance from the library, $\boldsymbol{p}$ is its projection on the column space of the data matrix D. The method based on the already mentioned fact that, in the absence of experimental errors, the vector-spectrum of the proper component should lie in the AOD hyperplane formed by the spectra of mixtures (see *Figure 1.1*). Consequently, to verify the number of components, it is necessary to organize a numerical evaluation of the degree of proximity of the reference spectrum to the subspace of mixtures.

The subspace dimension does not substantially affect the similarity criterion value. For this, one proposes to use the cosine value

$$\cos(\boldsymbol{t}, \boldsymbol{p}) = \frac{\sum_{i=1}^{N} t_i \cdot p_i}{\left[\sum_{i=1}^{N} t_i^2 \cdot \sum_{i=1}^{N} p_i^2\right]^{1/2}}, \tag{1.39}$$

or the correlation coefficient

$$r(\boldsymbol{t}, \boldsymbol{p}) = \frac{N\sum_{i=1}^{N} t_i \cdot p_i - \sum_{i=1}^{N} t_i \cdot \sum_{i=1}^{N} p_i}{\left\{\left[N\sum_{i=1}^{N} t_i^2 - \left(\sum_{i=1}^{N} t_i\right)^2\right] \cdot \left[N\sum_{i=1}^{N} p_i^2 - \left(\sum_{i=1}^{N} p_i\right)^2\right]\right\}^{1/2}} \tag{1.40}$$

The most difficult is the choice of the threshold value of the criterion of the component presence in the mixture. The practice of solving model problems showed that for IR spectra of organic compounds, the angle = 8 – 10°, and $r = 0.92 – 0.95$ could be taken as threshold criterion values. We do not recommend visual comparing $\boldsymbol{t}$ and $\boldsymbol{p}$ contours because similar curves can give completely unsatisfactory results. Similar to the previous method, nonparametric criteria for autocorrelation and randomness applied to the vector of difference between the reference and the projection can be used. Table 1.4 shows the results of determining the number of components for the set of spectra No. 2 (*Figure 1.4*) when projecting known spectra of each component. The criterion of the value of the angle between the reference spectrum vector $\boldsymbol{t}$ (*Figure 1.1*) and its projection onto the matrix column space $D^{\#}$, reconstructed using equations (1.8-1.12), was applied in this case for a different number of singular vectors, starting from 3. An abrupt decrease in the angle value corresponds to the number of base vectors equal to the number of components.

A known component can be artificially added to the mixture sets and used as a reference when using this method.

Table 1.4. **Projection of known component spectra onto the column subspace of the data matrix of 6-component mixture No. 2. The angle values (in degrees) between the vector of a known component $\boldsymbol{t}$ and its projection $\boldsymbol{p}$ onto the matrix $\boldsymbol{D}^{\#}$, restored using hypotheses $K = 3,4, ..., 10$. The jump in the angle value (decrease) corresponds to the number of components decreased by 1. The estimate equal to 6 coincides with the correct number of principal components.**

Hypothesis about the number of components	10	8	6	5	4	3
Component						
isooctane	1.52	1.53	**1.95**	5.7	14.5	20.0
benzene	6.1	6.5	**8.5**	28.7	41.1	44.8
toluene	8.1	9.0	**10.0**	41.0	45.3	53.8
diethyl ether	1.93	2.54	**3.0**	8.83	20.0	21.8
acetone	2.83	3.62	**5.1**	10.4	31.3	48.0
amyl acetate	0.81	0.82	1.60	**2.70**	9.73	17.3

In addition to the mentioned similarity criteria, in (Rasmussen, 1979), three groups of similar comparison procedures that give satisfactory practice results were proposed. The first consists of an explicit calculation of the distance between the vectors $\boldsymbol{t}$ and $\boldsymbol{p}$ (Δ in *Figure 1.1*), for example, by the Gram-Schmidt orthogonalization method. This distance essentially represents the magnitude of the residuals for an incompatible system of linear equations $D\boldsymbol{x} = \boldsymbol{p} \approx \boldsymbol{t}$. The second method, Target Transformation Analysis (TFA), is similar to the first, except that the matrix D is approximated by a matrix of the rank equal to the expected number of components. The third method (Bessel's inequality test) calculates the projection length $\boldsymbol{p}$ provided that $\boldsymbol{t}$ is normalized to 1. The similarity coefficient, b, is calculated as the sum of scalar products of $\boldsymbol{t}$ with K first singular vectors D:

$$b = \sum_{k=1}^{K} \left(\boldsymbol{t}^T \cdot \boldsymbol{u}_k\right)^2 . \tag{1.41}$$

It takes values from 0 to 1 and is essentially equal to the square of the projection length $\boldsymbol{p}$. A value of b close to unity is evidence of a good fit to the data.

The Durbin-Watson autocorrelation criterion (1.21) may also be applied to the elements of the difference vector.

The considered estimates of the validity of the hypothesis about the presence of an individual spectrum in a set of mixture spectra can be used for the data analysis by the library search method. The reliability of the solution, in this case, depends on the completeness of the spectroscopic database.

References

Akbari, M., Abdollahi, H. (2013). *J. Chemometrics*, **27**, 278-286.

Astakhov, S.A., Stögbauer, H., Kraskov, A., Grassberger, P. (2006). *Anal. Chem.*, **78**, 1620–1627.

Bu, D., Brown, C. W. (2000). *Appl. Spectr.*, **54**, No. 8, 1214-1221.

Bulmer, J. T., Shurvell, H. F. (1973). *The Journal of Physical Chemistry*, **77**, No. 2, 256-262.

Culler, S. R., Gillette, P. C., Ishida, H., Koenig, J. L. (1984). *Appl. Spectr.* **38**, No. 4, 495-500.

Durbin, J., Watson, G. S. (1971). *Biometrika,* **58**, 1–19.

Fok, M.V. (1973). Separation of Complex Spectra into Individual Bands by the Generalized Alentsev Method, *in: The Lebedev Physics Institute Series. Luminescence and Nonlinear Optics. Ed. by D. V. Skobel'tsyn.* Springer-Verlag US, Boston, MA, pp.1-22.

Friedrich, H.B., Yu, J.-P. (1987). *Appl. Spectr.*, **47**, No. 2, 227-234.

Gillette, P. C., Lando, J. B., Coenig, J. L. (1982). *Appl. Spectr*, **36**, No. 6, 661–665.

Hirschfeld, T. (1976). *Anal. Chem.*, **48**, No. 4, 721–723.

Hyvarinen, A., Karhunen, J., Oja, E. (2001) *Independent Component Analysis*, New York: John Wiley & Sons.

Johnson, N. L., Leone, F. C. (1977). *Statistics and Experimental Design in Engineering and the Physical Sciences*, New York, John Wiley & Sons.

Kawata, S., Komeda, H., Sasaki, K., Minami S. (1985). *Appl. Spectr.*, **39**, No. 4, 610-614.

Keller, H. R., Massart, D. L. (1992). *Chemometrics and Intelligent Laboratory Systems*, **12**, 209-224.

Knorr, F. J., Harris, J. M. (1981). *Anal. Chem.*, **53**, No. 2, 272-276.

Lancaster, P., Tismenetsky, M. (1985). The Theory of Matrices: With Applications, Academic Press; 2nd edition.

Lawley, D. N. (1971). *Factor analysis as a statistical method*, London, Butterworths.

Lawton, W.H., Sylwestre, E.A. (1971). *Technometrix*, **13**,. No. 3, 617-633.

Lee, T.-W. (1998). *Independent component analysis: Theory and applications*, Boston: Kluwer Academic Publishers.

Leger, M. N., Wentzell, P. D. (2002). *Chemometrics and Intelligent Laboratory Systems*, **62**, 171–188.
Lorber, A. (1984). *Anal. Chem.*, **56**, No. 6, 1004–1010.
Malinowski, E. R. (1981). *Anal. Chim. Acta.*, **133**, 99-101.
Malinowski, E.R., Howery, D.G. (1980). *Factor Analysis in Chemistry* New York: R.E. Krieger Publishing Company.
Martens, H. (1979). *Anal. Chim. Acta*, **112**, 423-442.
Mas, S., Tauler, R., de Juan, A. (2011). *Journal of Chromatography A*, **1218**, No. 51, 9260-9268.
Meisburger, S. P., Ando, Da Xu, N. (2021). *IUCrJ*, **8**, Part 2, 225-237.
Meister, A. (1984). *Anal. Chim. Acta*, **161**, 149-161.
Ohta, N. (1973) *Anal. Chem.*, **45**, No. 3, 553 – 557.
Rajko, R., Istvan, K. (2005). *J. Chemometrics*, **19**, 448–463.
Rajko, R. (2006). *J. Chemometrics*; **20**, 164–169.
Rodionova, O.Ye., Pomerantsev, A.L. (2005). *Progress in Chemometrics Research*, New York: NovaScience Publishers.
Sawal, M., Kubis, C., Franke, R., *et al.* (2014). *ACS Catalysis*, **4**, 2836-2843.
Shrager, R. I., Hendler, R. W. (1982). Anal. Chem., **54**, No. 7, 1147-1152.
Stogbauer, H., Kraskov, A., Astakhov, S. A., Grassberger P. (2004) *Phys. Rev.*, **70**, 066123-066140.
Volkov, V. V. (1996). *Applied Spectroscopy*, **50**, No. 3, 320 – 326.
Wald, A., Wolfowitz, J. (1940). *The Annals of Mathematical Statistics*, **11**, 147–162.
Wilcoxon, F. (1945) *Biometrics Bulletin*, **1**, 80-83.

CHAPTER TWO

SEPARATION OF MULTIVARIATE ADDITIVE SPECTRA INTO PURE COMPONENT SPECTRA

2.1. Introduction

Calculation of the spectra of individual components and their concentrations only from experimentally obtained mixed spectra represents one of the essential directions of chemometrics. In many cases, this approach turns out to be the only available way to determine the signal contours of components in mixtures of unknown compositions without using a physicochemical model or a priori information about the system.

This Chapter considers relatively easy-to-use separation methods that use solution filters such as non-negativity of components and their corresponding concentrations together with requirements of the simplicity of spectral shape and a maximum difference of individual contours. For such a problem, we will consider the singularity conditions of the solution. The methods under consideration are universal, but we will demonstrate their work on the example of the separation of infrared spectra.

For better understanding, a reader is advised to read the basics of the method outlined in Section 1.2.

2.2. On the uniqueness of the solution of the decomposition spectrum problem

Let us consider some aspects of the decomposition problem in more detail. The main question in the analysis of mixed signals is what properties of the shape of the contours of the components and/or concentration profiles can uniquely provide their finding. There is at least one such property: the presence in the component spectra of regions of mutual nonoverlap, arranged in such a way that **in any pair of the <u>mixture</u> contours, at least two continuous regions should present, whose nonzero intensity is due to only one component unique for each region** (Alentsev-Fok conditions

(Fok, 1973)). This condition is very hard in practice, and we will show later that it can be weakened considerably. For the Alentsev-Fok conditions to be provided, component spectra must contain nonoverlapped parts. In *Figure 1.1*, vectors $\boldsymbol{e}_1^{AF}$ and $\boldsymbol{e}_2^{AF}$ have zero intensities at λ_1 and λ_2, correspondingly and could meet the specified conditions of nonoverlap if they included several points, each of which lies in the nonoverlapped segments.

This vector representation in *Figure 1.1* demonstrates **the condition of the uniqueness of the solution of the decomposition problem: it is necessary and sufficient** that the nonnegative spectra of the components have nonoverlapped **points** at different abscissa values. Thus, we can formulate the following condition for the uniqueness of the decomposition (1.2), which is significantly weaker than the Alentsev-Fok rule: **in each pair of the component spectra, there must be at least two points, in one of which only the first component has a nonzero intensity, in the other only the second one.** The set of the component spectra is the set of vectors representing the intersection lines of the hyperplane formed by the vectors of mixture spectra (in the following, we will call it the mixture hyperplane) with the coordinate planes.

The intersection uniqueness proves the necessity of the condition. The existence of an exact procedure for finding an intersection proves the sufficiency.

Based on this fact, it is possible to suggest several numerical methods that use the information on the presence (but not necessarily the location) of null segments or even single points in the spectra of components.

2.3. Iterative projection method

The spectrum of a single component can be found using the projection procedure, by setting one or more nonzero points in the testing spectrum $\boldsymbol{t}_1$ (*Figure 1.1*), corresponding to the position of the most intense peak in the mixture spectra, and projecting $\boldsymbol{t}_1$ to a subspace of matrix columns using the transformation

$$\boldsymbol{p}_1 = U \cdot U^T \cdot \boldsymbol{t}_1, \tag{2.1}$$

where $\boldsymbol{p}_1$ is the projection vector. The matrix U here is the matrix of the left singular vectors from the expansion (1.8). The projection operation provides one of the conditions for admissibility of solutions – projection spectra: they are linear combinations of mixture spectra. This condition and the presence of correlations structure-spectrum do not allow one to construct the component spectra from the spectral bands available in the mixture spectra arbitrarily.

As mentioned above, at the first stage, an artificially prepared spectrum, consisting of one or more intense bands known to be contained in the individual contour, is taken as the initial approximation $\boldsymbol{t}_1$. The spectral peak relative intensities, shape, and position can not be specified precisely. It is desirable not to include "alien" intense bands, as it can sometimes complicate the further procedure. In the case, for example, the IR spectra of organic compounds, one may take a spectrum consisting of only one band as a starting approximation.

The obtained estimate $\boldsymbol{p}$ is refined in the iteration process, which consists in successive construction of the next trial vector $\boldsymbol{t}_2$ from $\boldsymbol{p}_1$, obtained at the previous step, by zeroing out all negative areas and bands decreased after the projection (we can assume that they are "alien"). Second, it is possible to enhance the newly appeared bands in the projection spectrum, as they may belong to the component but not be considered at the previous step. It is recommended to zero all areas, except for the second-most intense peak. After that, the obtained contour is projected again, and the changes are noted. The latter can be done visually or with the help of some numerical criterion of comparison of two curves (correlation coefficient (1.40), cosine of the angle between vectors (1.39), *etc.*).

The procedure of building the projection spectrum and its modification is repeated until the vectors $\boldsymbol{i}_n$ and $\boldsymbol{p}_n$ stop changing in the main details (if there is no noise, their shapes should coincide with the actual spectra of the components). When the fixation of a group of successively found bands does not cause new ones to appear, it is necessary to alternately exclude both the high and low-intensity bands from the obtained contour using a similar sequence of operations. If this sequence does not give more "foreign" bands in the spectrum found (the excluded intensities are restored after projection), the final projection spectrum is considered the pure component spectrum.

When all bands in the mixture spectra are exhausted, all the component spectra (matrix E) are found, the corresponding concentration matrix C is calculated using (1.2). The negativity of elements of this matrix indicates the incorrect shape of individual spectra, so they should be corrected by zeroing the negative values in C and calculating the corresponding component matrix E followed by using the resultant spectra projection. The obtained individual spectra reproduce (within the errors of the initial data, if the number of basis vectors in U is less than the number of mixtures) the whole set of the mixture spectra and, hence, satisfy the decomposition (1.2).

Figure 2.1 shows the iterative design results for the three components of the 5-component set of IR mixture spectra **No. 2** (*Figure 1.4*).

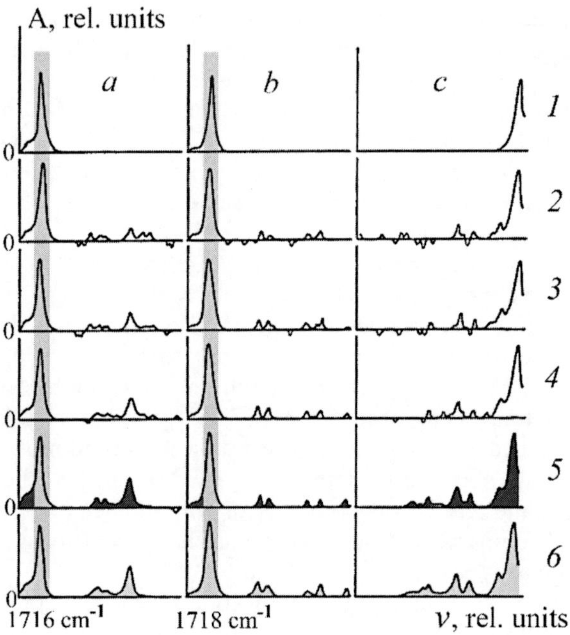

Figure 2.1. Steps of reconstruction using target transformation of component spectra from the 5-component mixture data (set **No. 1**). Spectra of acetone (a), cyclohexanone (b), and diethyl ether (c). 1 – initial spectrum t1, 2 ÷ 5 spectra of $\boldsymbol{p}_{1-4}$ projections for successive projection steps, 6 – component spectra measured for pure substances. The grey areas mark peaks overlapped in components *a* and *b*. The positions of their maxima are shown below the abscissa axis.

Looking at the spectra of the mixtures, we could see that the first left intensive absorption band (at 1717 cm^{-1}) changed the position of its maximum from mixture to mixture from 1718 to 1716 cm^{-1} and the shape of its wings changed in the process. This fact suggested that this band was a superposition of two bands belonging to different components. Indeed, *Figure 2.1a,b* shows that if a contour consisting only of the 1718 cm^{-1} band or only of the 1716 cm^{-1} band is given as the initial spectrum, one can find the two spectra that differ in other regions – the spectra of acetone and cyclohexanone. It should be noted that isolation of the spectrum of cyclohexanone was not too difficult, while for acetone, the convergence of the method was much worse until the correct band shape at 1718 cm^{-1} was achieved. The difficulty of finding these spectra was that the characteristic spectral features were significantly overlapped, and the remaining areas were low-intensity and overlapped with other components. It was not difficult to isolate the remaining spectra since they were sufficiently distinct.

The program implementing the described method showed that the dialogue mode of operation is the most effective. The operator makes corrections to the spectra by observing them on display. It is difficult to program an automatic mode of operation for all cases since it is necessary to create convenient and universal algorithms for recognizing and analyzing spectral images that formalize the correction processes. On the other hand, the dialogue mode allows the expert to quickly and visually use all the available spectral information about the solving problem.

2.4. Self-modeling method for searching spectra of components using additional information and constraints

There are problems where even accurate initial information does not provide separation to the proper spectra of components. The main peculiarity is a set of permissible solution variants that formally satisfy the non-negativity of matrices E and C. The selection of one solution from this set requires a variety of additional information, which is often difficult to formalize in the framework of the decomposition method used. In turn, it can lead to subjectivity in evaluating individual spectra. It should be emphasized that subjectivity does not consist in the complete arbitrariness of the choice of a solution but in the subjectivity of the importance of additional information, which will take the form of weighting coefficients. This section will consider a search method belonging to the self-modeling group, based on a singular analysis of the data matrix. This approach allows one to use various additional information about the problem and the requirements for the solution.

Most methods for solving spectral decomposition problems use different numerical procedures, usually adapted to the specific type of additional information used. With all the differences in their approaches, they are united by the fact that component search is conducted in the areas of admissible solutions of the linear problem (AOB and COD sub-cones in *Figure 1.1*). The search for spectra is based on the construction of linear combinations of vectors forming some basis of the subspace of mixture spectra. The orthonormal set of the left singular vectors U can be considered a basis.

Consider the approach based on the construction of a sequence of tentative spectra of components

$$x_{ik} = \sum_{j=1}^{L} u_{ij} \cdot y_{jk}, \quad K \le L \le M, , \tag{2.2}$$

where coefficients y_{jk} are the search parameters and arguments of some target function $W(y)$, which at the minimum satisfy the used constraints and conditions reflecting various additional information. K is the number of components, M is the number of mixtures, and L is the number of singular vectors $\boldsymbol{u}$ in the sum (2.2) (remember that singular vectors are ordered according to the decreasing order of singular values, that is, only significant vectors are used).

The peculiarity of the proposed method is that there is not necessary to include in the target function the discrepancy value between the model spectra calculated by the relation

$$D^{\#} = X_n C_n, , \tag{2.3}$$

where n is the number of the step (or the search iteration), and the experimental data:

$$\text{RES} = \frac{1}{N \cdot M - K} \left\| D_{эксп} - D_n \right\|^2 \tag{2.4}$$

The discrepancy RES is determined by the remainder matrix F^0 in $D = D^{\#} + F^0$ (1.13), which, in turn, is equal to the product of singular matrices: $F^0 = U_{M-L} \Lambda_{M-L} V_{M-L}^T$. The index $M - L$ indicates that the last $M - L$ columns are left in the matrices, and L equals the number of the first singular vectors taken for constructing the model matrix of components. Thus, whatever spectra of the components are calculated in the search process, they will lie strictly in the subspace $D^{\#}$, and formally satisfy the condition of the solution. However, the restrictions imposed on the shape of the component contours increase the minimum discrepancy, which can be estimated from the remainder matrix calculated using singular vectors with indices L+1, ..., M, which are not included in the number of vectors, from which the component spectra are constructed:

$$\text{RES}_{\text{lim}} = \frac{1}{N \cdot M - L - 1} \left\| R \right\|^2, \quad R_{ik} = \sum_{j=L+1}^{M} u_{ij} \cdot y_{jk} \,. \tag{2.5}$$

Inclusion of this residual in the target function is not necessary in case of non-overlapping conditions, but, as the practice has shown, it accelerates the search for the minimum of the target function in the case of large size of the problem, when the number of components is more than 4-5. A large number of penalty terms in the target function might make it multimodal, *i.e.*, a set of local minima may exist corresponding to different sets of component spectra corresponding to a group of conditions imposed on them. In this case, the inclusion of the residual term improves the convergence of the method.

Thus, the target function W, proposed in (Volkov, 1996) and realized in the program REMIX, may be extended to the following form:

$$W(\boldsymbol{y}) = \alpha \cdot (\mathrm{RES} - \mathrm{RES}_{\mathrm{lim}}) + \beta \cdot [\mathrm{cond}(X) - 1]^2 + \gamma \cdot \frac{1}{N \cdot K} \cdot \sum_{\forall (x_{ik} < 0)}^{N,K} x_{ik}^2 +$$

$$+ \delta \cdot \frac{1}{M \cdot K} \cdot \sum_{\forall (c_{k,m} < 0)}^{K,M} c_{ik}^2 + \varepsilon \cdot \frac{1}{N \cdot K} \cdot \sum_{k=1}^{K} len\,(\boldsymbol{x}_k) + \qquad (2.6)$$

$$+ \eta \cdot \frac{1}{N \cdot K} \cdot \sum_{k=1}^{K} cur\,(\boldsymbol{x}_k) + \xi \cdot \frac{2}{K \cdot (K-1)} \cdot \sum_{\forall (l \neq m)}^{K} [\cos(\boldsymbol{x}_l, \boldsymbol{x}_m)] +$$

$$+ \tau \cdot \frac{2}{K \cdot (K-1)} \cdot \sum_{\forall (l \neq m)}^{K} r\,(\boldsymbol{x}_l, \boldsymbol{x}_m) + \ldots ,$$

where X is the matrix of tentative spectra of the components; $\boldsymbol{x}_l$ and $\boldsymbol{x}_m$ are the lth and mth columns of this matrix; $i = 1, ..., N$ is the index of the point in the spectrum; $k = 1, ..., K$ is the index of the component; r is the pairwise correlation coefficient between the spectra. *len*() and *cur*() are vector length and curvature of the spectrum contour, correspondingly.

Let us consider the second term in more detail. Here cond(X) denotes the condition number of the matrix of tentative normalized spectra of components X calculated as the ratio of the maximum to the minimum singular values of X, i.e., the diagonal elements of the matrix from the expansion

$$X = U_x \Lambda_x V_x^T, \quad \mathrm{cond}(X) = \frac{\lambda_{\max}}{\lambda_{\min}}. \qquad (2.7)$$

The condition number, in this case, reflects the total similarity of the vector spectra of the components, reaching unity at the moment of their maximum difference (in this case, being orthogonal, the component vectors must contain the negative elements or coincide with the coordinate axes). Subtracting unity from the similarity criterion is due to this fact. Geometrically, small cond(X) values correspond to the large cone angle formed by the component vectors $\boldsymbol{x}_j$.

The third and fourth terms in (2.6) are penalties for the negative elements in the model spectra and the negativity of the concentration matrix elements defined by relation (2.3).

The fifth term is the normalized spectral contour length approximated by a broken curve

$$len(\boldsymbol{x}_k) = \sum_{i=1}^{N-1} [(\boldsymbol{x}_{i+1,k} - \boldsymbol{x}_{ik})^2]^{1/2}. \qquad (2.8)$$

The expression for curve length acts as a penalty for the complexity of the spectral contour. As already mentioned, the spectral contour is normalized before calculating its length by dividing it by the maximum intensity value. It seems more reasonable to use normalizations to the unit Euclidean norm, or norm 1 (the sum of the element modules). However, numerical experiments have shown that the sensitivity of the target function to the overlap of the component contours is worse with this scaling.

The remaining terms in (2.6) are redundant or alternative to the related terms considered above. Thus, the sixth term contains the expression

$$cur(\boldsymbol{x}_k) = \sum_{i=2}^{N-1} \left| (\boldsymbol{x}''_{ik}) \right| , \tag{2.9}$$

which is proportional to the curvature of the spectral contour, calculated as the sum over the second derivatives of the vector-spectrum normalized by the maximum element. This criterion is considered as a criterion of spectrum complexity. The practice of solving models and real problems has shown that the inclusion of this term in (2.6) in rare cases provides the smooth spectrum contours, but, as a rule, its use worsens the conditionality of the problem and prevents obtaining an acceptable solution.

The second term reflects an integral characteristic for all components. The seventh and eighth terms in (2.6) work similarly. Their minimization leads to an increase in the angles between the component vectors. Since these terms express the same requirement, their simultaneous use is hardly advisable.

Introducing several essentially similar penalty terms into the target function is helpful in practice since such terms have different sensitivity to details of spectral contours, and the researcher has an opportunity to select appropriate weight coefficients for the particular class of problems to be solved and to achieve the most efficient operation of the algorithm.

Another criterion for similarity (or difference) of spectra, sometimes used by other authors in their works, is the polyhedron volume formed by the ends of the specially normalized vector-spectra of components (Meister, 1984). The application of this criterion significantly complicates the search for a solution as the number of components increases, since it requires extensive calculations for constructing a polyhedron with the maximum possible volume (for this purpose, (Meister,1984) generates all possible trial nonnegative spectra of components with a reasonably small step and selects the best option).

In the literature on statistics, several similarity criteria for objects are known. These are the Euclidean metric defined by the square of the norm of the difference between two vectors, the Manhattan metric $\sum_{i=1}^{N} |a_i - b_i|$,

the Bray-Curtis metric $\dfrac{\sum_{i=1}^{N} |a_i - b_i|}{\sum_{i=1}^{N} a_i - \sum_{i=1}^{N} b_i}$, Canberra distance $\sum_{i=1}^{N} \dfrac{|a_i - b_i|}{a_i - b_i}$, and some others, including the value of information entropy. Verification of the corresponding expressions has shown that they significantly degrade the properties of the target function in the vicinity of the solution due to their low sensitivity to small changes in the shape of spectral contours.

Generally speaking, it is possible to propose various criteria of complexity or simplicity of spectra. However, it is difficult to present sufficiently weighty grounds for their use since the connection of such criteria with the physical meaning and features of particular problems is very indirect and difficult to formalize. The practice of solving model and real decomposition problems has shown that the value of information entropy (1.6), as a complexity criterion for UV, visible, IR, and mass spectra, is very unrepresentative. It cannot provide an acceptable convergence rate of the minimization procedure (at least, for the number of components more than two, which may explain the fact that the known examples of decomposition are not more than 2-3 component problems), so the criterion (1.6) can be refused.

Some of the terms in (2.6) have a quadratic form. As noted in the literature on nonlinear minimization methods (Dennis, 1983), this should provide an acceptable rate of convergence of the solution search procedure when using quasi-Newton algorithms. However, starting from the fifth term, their quadratic forms worsen the conditionality of the problem in the vicinity of the minimum and increase the scatter of solutions at small variations of the weight coefficients at the penalty terms.

The functional (2.6) minimization means searching for a given number *K* of individual spectra. These spectra lie in the subspace of the mixture spectra and, therefore, can formally be spectra of components. At the same time, the conditions of maximum distinction and simplicity of the spectral contours are met.

Thus, expression (2.6) is a penalty function. As is known (Gill, 1981), the method of the penalty functions is far from being optimal from the point of view of stability of the solution of the problem because the condition number of the matrix of second derivatives of the target function often turns out to be so large that the numerical procedure of finding the minimum is sometimes not able to reach the neighborhood of the solution. Nevertheless, this approach allows one to extend the target function simply by adding new penalty terms, considering any additional information, if it can be formalized in the form of a quadratic expression.

The problem conditionality may be improved by using known spectra of some components as additional information. In this case, we can include them in the matrix *Xn* and fix them, forbidding their change. This procedure reduces the number of unknowns. The *a priori* knew spectra should be included in the component matrix with high accuracy. Even minor systematic errors in the intensity under poor conditionality can lead to significant distortions of the searched spectra, up to the impossibility of identifying them. If there is no confidence in their accuracy, one may recommend not to use the known spectra when solving the decomposition problem.

Experience of using the main known minimization methods of zero-order (without calculation of first derivatives of the target function) and first-order (using the finite-difference first derivatives) was shown that the most stable are the first-order Broyden-Fletcher-Goldfarb-Shanno variable metric scheme (Gill,1 981; Dennis, 1983) and the zero-order Nelder-Mead polyhedron algorithm. They are used in the REMIX decomposition program (Volkov, 1996), which was implemented in the analysis of 6-component mixture spectra. In this program, the values of the weight coefficients are set automatically (the mixture spectra are normalized to the unity maximum): $\alpha = 2.0$, $\beta = 1.0$, $\gamma = 100$, $\delta = 103$, $\varepsilon = 100$, $\eta = 5.0$, $\xi = 2.0$, $\tau = 100$. Coefficient values are set from considerations of approximate equality of the target function terms at the minimum point, except for the first term, which contribution is sufficient at the level of ~0.01 from contributions of other terms. It is desirable to keep the contribution of terms (2.7) and (2.8) at the level of 10-30. Practice shows that the convergence rate with this choice is quite high in most cases, and the solution meets the chosen requirements.

We will demonstrate the operation of the REMIX algorithm using the example of the analysis of a 6-component mixture of organic substances (data set No. 2 from Section 1.4.1). The set of experimental spectra is shown in *Figure 1.4.* The estimation of the number of components is discussed in Sections 1.4.2 and 1.4.3. *Figure 2.2* shows the solutions under the hypotheses $K = 6, 7, 8$ and the number of singular vectors in (2.2) $L = 6, 7, 8$, respectively.

This experiment demonstrates that increasing the number of model components does not lead to significant distortions of main spectral contours $1 \div 6$. At the same time, the introduction of additional components makes it possible to remove some spectral bands (caused by additivity disturbances due to the interaction of the components with each other) from the model components from the bands.

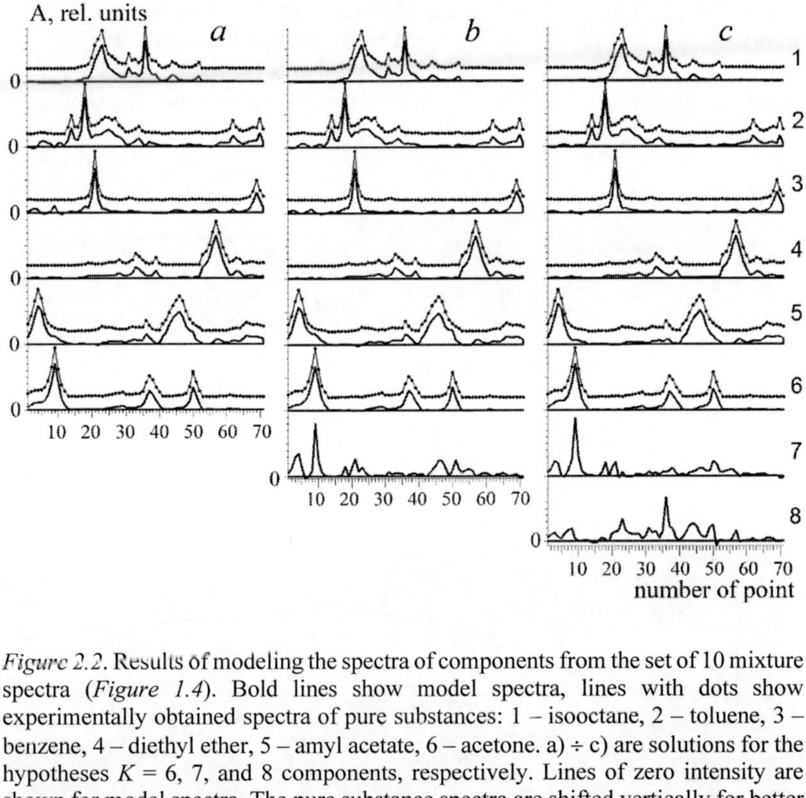

Figure 2.2. Results of modeling the spectra of components from the set of 10 mixture spectra (*Figure 1.4*). Bold lines show model spectra, lines with dots show experimentally obtained spectra of pure substances: 1 – isooctane, 2 – toluene, 3 – benzene, 4 – diethyl ether, 5 – amyl acetate, 6 – acetone. a) ÷ c) are solutions for the hypotheses $K = 6$, 7, and 8 components, respectively. Lines of zero intensity are shown for model spectra. The pure substance spectra are shifted vertically for better visualization.

In particular, on the example of components 2 and 3, one could see that at $K = 6$ (solution *a*), the small peak in the region of points 1-15 decreases, appearing in component 7 (solutions *b*,*c*). This peak belongs to the solvent spectrum (carbon tetrachloride) incompletely compensated in the mixture spectra. Consequently, the estimate of the number of components 7 (Section 1.4.3) turns out to be closer to the truth. At $K = 8$, spectrum No. 8 cannot be identified because of the large number of bands, which correlates with the band in spectrum 1. One may propose that this component spectrum includes other impurities in the isooctane used to prepare the model mixture.

When the number of components is overestimated (here: 8), the search program is forced to construct additional nonnegative contours with nonnegative contributions in the mixture spectra. These contours can represent sums of the contours of the other components plus the weighted "noise" singular vector No. 8, participating in the construction of the model matrix X, and its interpretation should be performed with caution.

Distributions of concentrations of components for the solution $K = 7$ are shown in *Figure 2.3*. One can see that the contribution of the 7th component is tiny.

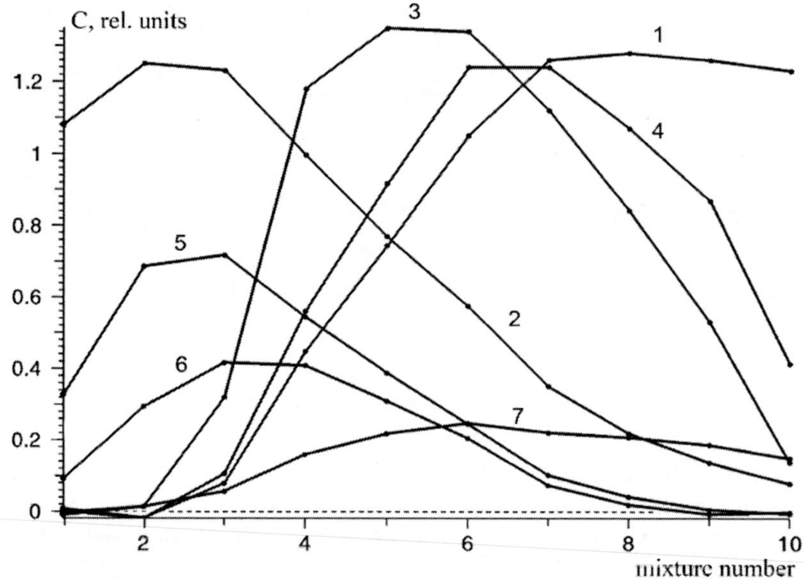

Figure 2.3. The component concentrations for solution (*b*). Numbers along the abscissa axis correspond to *Figure 1.13*.

Residual matrices $F_K^\# = D - X_K \cdot C_K$ at $K = 5, 6, 7$, and 8, presented as contours made up of columns $F_K^\#$, are shown in *Figure 2.4.* As the number of components in the model increases, the amplitude of the residual matrix elements decreases. The Wald-Wolfowitz criterion (1.38) for the hypothesis of the randomness of the residual matrix elements does not show the absence of correlations. It cannot provide a reliable determination of the number of components. The Durbin-Watson criterion (1.21) turns out to be less sensitive to small systematic deviations (which are presumably caused by intermolecular interactions), as shown in Table 4 and, with a confidence interval {1.7-2.2} shows 8 components. The example demonstrates a typical difficulty because the dispersion of measurement noise may vary significantly from mixture to mixture and make statistical analysis of the residue matrix dispersion value (assuming the same dispersion magnitude for all the spectra) inappropriate. Tests for the absence of autocorrelations in the residuals in such cases are mainly free from such restrictions.

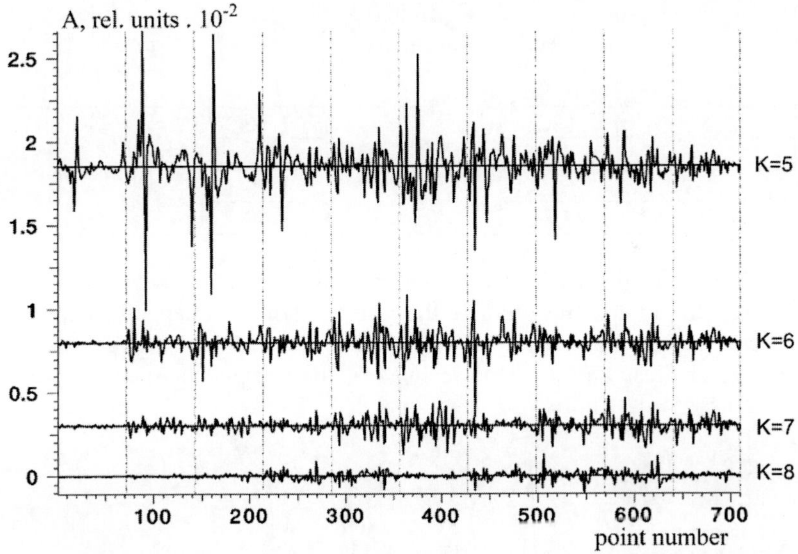

Figure 2.4. Residuals combined for the different components *K*. Dashed vertical lines separate the mixture spectra (71 points per spectrum).

***Table 2.1*. Autocorrelation and randomness criteria for the elements of the residue matrices at the different hypotheses of the number of components.**

Number of components K	Wald-Wolfowitz criterion (1.38)		Durbin-Watson criterion (1.21)	
	WW	Confidence interval at Q=0.05	DW	Confidence interval at Q=0.05
5	-12.2	-3.13, 3.13	1.16	1.65, 2.18
6	-6.38		1.56	
7	-6.23		1.59	
8	-7.35		**1.68**	

The artificial set of mixture spectra was prepared to demonstrate the peculiarities of decomposition in the case of wholly overlapped component spectra (*Figure 2.5*). The noise was not introduced into the model mixture spectra to illustrate the shape features of the component spectra without unnecessary distortions.

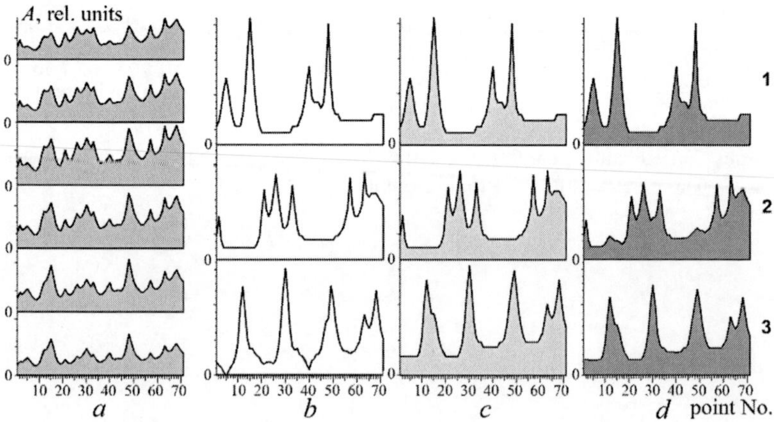

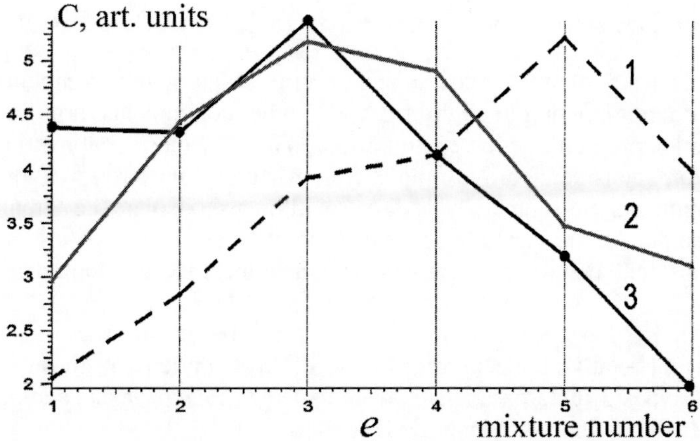

Figure 2.5. The artificial 3-component problem prepared from completely overlapped spectra of components. a) mixture spectra, b) solution at weights $\{\alpha = 1, \beta = 10, \gamma = 10, \delta = 10, \varepsilon = 10\}$, c) true component spectra, d) solution at weights $\{\alpha = 1, \beta = 10, \gamma = 10, \delta = 10, \varepsilon = 50\}$, e) the relative concentrations. The spectra are shifted vertically for better visualization. Numbers 1, 2, 3 indicate the numbers of components.

The solution was obtained using the first five terms in formula (2.6) and the weight coefficients given above. As is seen, the spectrum of the third component reaches zero near the 5th point (*Figure 2.5b*). Increasing the penalty for the spectral contour lengths by three leads to the correct form of the 3rd component, but the spectrum of the 2nd component turns out to be slightly distorted by the addition of the third contour. Thus, by selecting the relative values of the weights of the penalty terms, it is possible to influence the quality of the solution and investigate the mutual influence of the individual contours found. In complicated cases, one can increase the penalty weights for each component individually in the dialogue iterative procedure.

2.5. Analysis of data with violation of additivity

Analysis of the mixture spectra in which some components are compounds existing in equilibrium with other components turns out to be a complex problem, which could be solved only by numerical methods. This section considers an example of mixtures of strongly interacting components. Such an analysis does not allow us to isolate the actual spectra of the pure components but at least separate the sets of spectral bands into groups that behave as components when the mixture compositions are varied.

Figure 2.6 shows a fragment of the mixture spectra of DMFA at different relative concentrations of hexafluoroethylene. In these mixtures, three types of heteroassociates were expected to appear, depending on the ratio of hexafluoroethylene: DMFA = 1:1 ÷ 10:1.

The analysis of the singular vectors of the data matrix shows at least 11 components since the Durbin-Watson criterion (DW) did not exceed $0.7 < DW < 1.7$ for all sequences. Because the number of components is equal to the number of mixtures, we can assume that the number of associates is more than the expected $K = 3$, or intermolecular interactions make the system nonlinear. Singular vectors of the data matrix are shown in *Figure 2.7*. The shape of the contours contains information about changes in the shape of spectral bands during component interactions. *Figure 2.7b* shows that the contribution of the last three singular vectors to the data is relatively small, and the 11th vector practically corresponds to measurement noise. Decompositions into 10 and 8 components given in *Figure 2.8* show that the contours of the individual spectra are close between solutions and fit to **three distinct** characteristic shapes (spectra **2**, **3**, and **6** in the 8-component solution). The remaining components contain features related to shifts of the intensive bands due to the component interactions. One of the contours (4th) corresponds to an uncompensated background. Interestingly, the 6th component of the 8-component solution appeared together with the 7th and 8th contours. Further splitting to the 10-component solution shows this behavior.

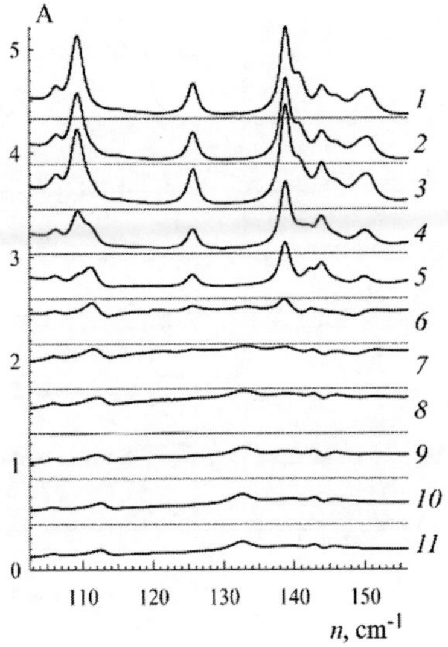

Figure 2.6. The IR mixture spectra of hexafluoroethylene with DMFA. For each spectrum, a zero absorption line is indicated.

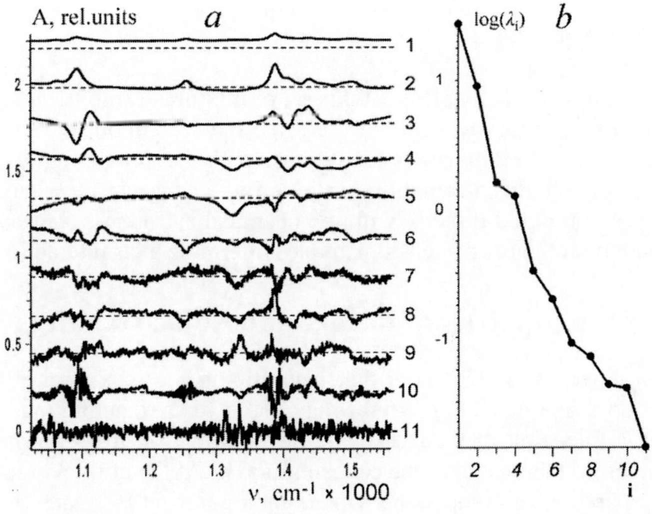

Figure 2.7. Singular vectors of the data matrix D in *Figure 2.6*. On the right is the spectrum of logarithms of the singular values of D; index i relates to their ordinal numbers.

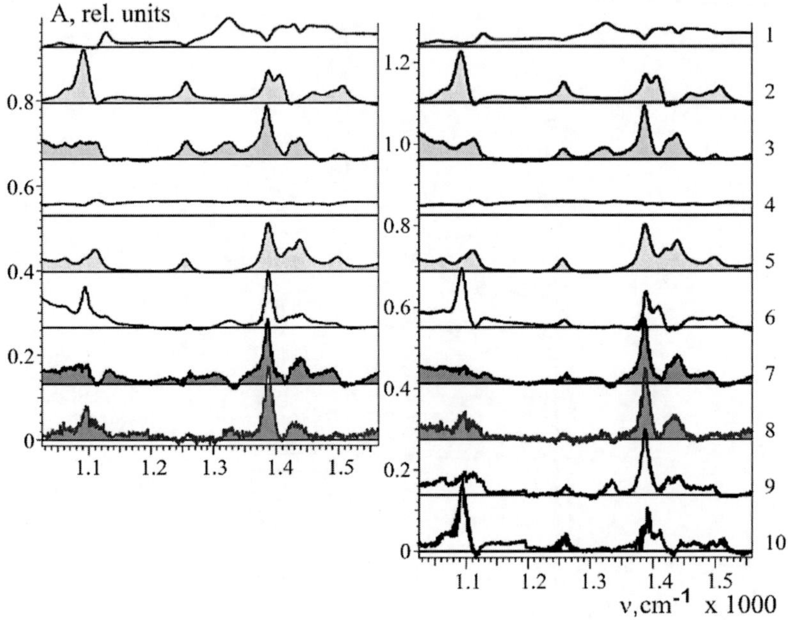

Figure 2.8. Decomposition of the IR spectra in *Figure 1.4* into eight (left) and ten components (right).

An example of the analysis of this set of mixtures demonstrates that the shape of the absorption bands of the three main components varies with changes in their relative concentrations, and the remaining calculated components contain information on the group change in the shape of the bands, not correlated directly with the chemical substances, but providing information on the presence and nature of intermolecular interactions.

2.6. Analysis of the decomposition stability

The above examples show that determining the maximum allowable number of components in mixtures may not have an unambiguous answer. Factors influencing the success of the solution of the decomposition problem are: 1) the value of the concentration matrix condition number (the better separation of components when obtaining a set of related mixtures, the smaller cond(C) and the better convergence of iterative search methods), 2) the matrix D condition number (the more dissimilar and nonoverlapped component spectra, the smaller cond(D) and the better all search methods work), 3) the spectral error level, 4) the search algorithm applied, *etc*. The condition number of a data matrix is determined by the ratio of its maximum and minimum singular numbers. The higher the number, the closer the matrix is to a degenerate one and the worse the decomposition stability.

For example, the authors (Fok, 1973) note that the Alentsev-Fok method permits the decomposition of mixed spectra of no more than 3-4 components. The original self-modeling method (Lawton, 1971) permits decomposition into the 2-3 components. The minimization of the target function (2.6) allowed one to quickly solve the decomposition problems of the 5-10 component mixtures. Apparently, the real boundary should be considered as 8-10 components with cond(D) = 50 – 100 and cond(C) = 10 – 50.

Evaluating the stability of the component spectra contours may be done by comparing the solutions obtained by varying the number of components under search, the number of singular vectors used in their construction, and comparing the solutions obtained using incomplete sets of experimental data. The latter method (the so-called "bootstrap" (Efron, 1979), or "jackknife" (Miller, 1974)) is based on constructing all possible combinations of M_t mixture spectra from M obtained in the experiment ($K \leq M_t < M$), and pairwise comparison of the spectra between the obtained solutions. Notice that this approach is efficient enough, whereas the methods based on the solution variance analysis work not always. In the latter case, the corresponding criteria yield the upper-bound estimates, leading to senselessly large practice values. In the next section, the peculiarities of the evaluation will be demonstrated by the example of solving the trivial problem of calculating the spectra of components with a known concentration matrix C.

2.7. Using known relative concentrations of components in mixtures

2.7.1. Basic principles

A priori knew values of the relative concentrations of the components in the mixtures reduce the analysis problem to a simple solution of a system of linear equations:

$$D = C \cdot X, \qquad C^T \cdot D = C^T \cdot C \cdot X \tag{2.10}$$

where D is the M by N matrix of mixture spectra, X is the K by N matrix of the desired component spectra. Hereafter, the notation "X" is used for the component spectra, and C is the M by K matrix of known relative concentrations.

Relative concentrations can be obtained, for example, as a result of analysis of known physical processes leading to changes in the contributions of components in mixtures. Estimates of C_{jk} elements are obtained from known experimental parameters such as time, initial concentrations, pH of the medium, temperature, *etc.*

In (Knorr, 1981), to decompose the fluorescence spectra of two-component mixtures of organic compounds (anthracene series), the matrix of relative concentrations was modeled as

$$C_{jk} = C_{jk}^{0} \cdot e^{\left(\frac{j \cdot \Delta t}{\tau_k}\right)}, \tag{2.11}$$

where Δt is the time interval elapsed from the beginning of the experiment, C_{jk}^{0} are the initial concentrations, and τ_k is the luminescence time. Parameters C_{jk}^{0} and τ_k are varied using an iterative minimization procedure. The spectra of components X calculated using equation (2.10) reproduce the mixture spectra in the best way, *i.e.*, minimize the value of the quadratic discrepancy.

If the values of the equilibrium constants are not determined, they are included in the number of unknowns. However, the decomposition problem solution becomes ambiguous (Sylvestre, 1974).

In the case of errors in the initial information, the system of equations (2.10) is inconsistent, leading to violations of additivity on the right-hand sides and in C. The most common method for solving such systems is the least-squares (LS) method. When the coefficient matrix turns out to be perturbed, this method gives biased and inconsistent estimates (Lawson, 1974). The problem (2.10) is a type of passive experiment. At the same time, the LS method assumes the case of an active experiment (when the elements of the matrix C are known precisely, and all the experimental errors are in the matrix D).

Various regularizing algorithms are used to solve ill-conditioned problems. Methods for constructing such algorithms for the case of a passive experiment are considered in (Voevodin, 1983). A general tactic for solving ill-conditioned systems of linear algebraic equations under conditions of initial data perturbation consists of the following steps. First, we obtain estimates of the Euclidean norms of experimental errors in the coefficient matrix:

$$\varepsilon_C = \left\| C - \widetilde{C} \right\|_E .$$

The difference is calculated between the exact and experimental matrices of coefficients, based on a priori information about the problem. Then the singular decomposition is calculated

$$\widetilde{C} = U \cdot \Lambda \cdot V^T \tag{2.12}$$

followed by the analysis of the values of the singular numbers (diagonal elements of Λ). Suppose that both inequalities proposed in (Voevodin, 1983) are fulfilled

$$\begin{aligned} &\left\|\Lambda^+\right\|_2 \cdot \left\|\Lambda\right\|_E \cdot \left(\varepsilon_C + 2.9 \cdot (M+K) \cdot p^{-t+1}\right) \le 0.1, \\ &\left\|\Lambda^+\right\|_E \cdot \left\|\Lambda\right\|_E \cdot \left(\varepsilon_C + 2.9 \cdot (M+K) \cdot p^{-t+1}\right) \le 0.1, \end{aligned} \tag{2.13}$$

where Λ^+ is a pseudo-inverse matrix, $\|\Lambda^+\|_2 = min\,\lambda_k$, λ_k is the singular number. The sum $\varepsilon_C + 2.9 \cdot (M+K) \cdot p^{-t+1}$ reflects the machine precision (p is the base of the arithmetic, t is the number of significant digits in the machine representation of real numbers, M is the number of rows, K is the number of columns of the matrix). The concentration matrix is considered in this case to have a full rank, and the problem (2.10) is parameterized by using additional information about the exact problem (for example, the value of the estimate ε). The system is solved by regularization methods, applying algorithms for finding nonnegative solutions.

The main problem of the decomposition with known estimates of concentrations, as in the search for components by other methods, remains the estimation of the solution reliability. Under a reliable solution, we will understand a set of spectra of components, for each of which we obtained individual estimates of the scatter of the solution.

Literature analysis has shown that insufficient attention has been paid to this aspect. It is desirable to use both an integral estimate of solution dispersion (which is usually so overestimated that it has no physical sense) and stability estimates for each component spectra. The integral estimate is given by the condition number of the coefficient matrix C, which sets an upper bound on the possible growth of the error rate δX in the solution as compared to the error rates in the initial data δC and δD (Strang, 1980; Voevodin, 1983):

$$\frac{\|\delta X\|}{\|X\|} \le cond(C) \cdot \frac{\|\delta C\|}{\|C\|}, \qquad \frac{\|\delta X\|}{\|X\|} \le cond(C) \cdot \frac{\|\delta D\|}{\|D\|}, \tag{2.14}$$

2.7.2. Determination of the component spectra by using Concentration estimates

Let us now consider the procedure for finding the spectra of components using estimates of their relative concentrations in more detail. Firstly, numerical simulations show that when solving ill-posed (ill-conditioned) problems, the methods of passive experiment do not allow us to obtain solutions with smaller error variance than the conventional LS scheme. Secondly, the magnitude of the shift in the LS solution at the typical experimental error levels turns out to be less than its actual variance. Thirdly, methods of passive experiment make it possible to obtain a small value of the bias only if the system $D = E \cdot C$ (1.2) is significantly overdetermined, which is not always feasible. Fourthly, these algorithms are usually iterative and require additional efforts to ensure their convergence in the neighborhood of the solution.

The practice of solving model problems has shown that methods of regularization allow obtaining spectra of components that are closer to the true ones in comparison with the conventional LS solution only in the case of very poorly conditioned problems (cond(C) > 500 at an error level δC in the matrix of the order of 1-3 % relative). It should be kept in mind that the regularization yields biased solutions, but the bias is usually less than the distortion of the shape of the spectra. If the cond(C) does not exceed 150-200, any parametrization, although it permits reducing the error in the solution for a single spectral point, does not provide small **relative** deviations between points in the spectrum, which are responsible for the distortion of the spectral contour. Since the shape of spectral curves in most cases is difficult to describe analytically, it is difficult to relate individual points in individual spectra to each other to form a target function that gives a solution with minimal error in the spectral images. *Figure 2.9* shows the decomposition of the data set **No. 2** considered in the previous section. The analytical chromatography data yields the cond(C) = 512. The parameterized solution corresponding to the equation

$$\widetilde{D} = \left(\widetilde{C} + \alpha E\right)\widetilde{X} \tag{2.15}$$

where $\alpha = {\varepsilon_C}/{\|\tilde{C}\|} = 0.03$ is the regularization parameter under condition $\|\tilde{C}\| = 1.0$, E is the unit matrix. Such regularization artificially reduces the condition number of the coefficient matrix and thus stabilizes the solution, increasing its bias but reducing the variance of the solution according to the estimates (1.61).

It turned out that the best results for problem (2.15) with matrices of any degree of conditionality can be obtained by solving it by weighted least squares

$$\left(\widetilde{C} \cdot W\right)^T \cdot \widetilde{D} \cdot W = \left(\widetilde{C} \cdot W\right)^T \cdot \widetilde{C} \cdot W \cdot \widetilde{X}\,, \tag{2.16}$$

in which the diagonal elements of the weight matrix w_{jj} for the equations $j = 1, ..., M$ are assigned equal to the inverse values of the total quadratic discrepancies for each of the equations after obtaining the primary solution by one of the methods discussed above:

$$w_j = \left[\sum_{k=1}^{K} \sum_{i=1}^{L} \left(\widetilde{X}_{ki} \cdot \widetilde{C}_{jk} - \widetilde{D}_{ji} \right)^2 \right]^{-1} . \tag{2.17}$$

Using the w_{jj} coefficients scaled by maximum value to 1, which form a diagonal matrix of weights, the new system of equations (2.15) is solved by the LS method. In practice, the more rows in the C matrix, the better this method works. In this case, if independent measurements of the mixture spectra are made, it is better not to combine them but to insert them into the system (2.15) independently of each other. Unfortunately, due to many sources of errors in the original information, often systematic, it is difficult to statistically justify a particular way of calculating the weights W, including a specific kind of formula (2.17).

Whichever method one uses to solve the decomposition problem, the resulting component spectra will be severely distorted in the case of an ill-conditioned C matrix.

The condition number cond(C) is a measure of the maximum growth of the error rate in the solution as compared to the error rate in the original information (equation 2.14) only under small perturbations, when the error matrix C_e in the representation

$$\widetilde{C} = C + C_e \tag{2.18}$$

has such a small norm that C remains the total rank for all realizations. From the one side, one cannot use the condition number to measure the solution instability (or unreliability) since this number is close to random. From the other side, practice shows that cond(C) can still serve as an empirical criterion for the overall reliability (or variability) of a solution since the discrepancy between the calculated and proper component spectra is increased with increasing cond(C).

In addition to a general estimation of the variability of the solution under perturbation of the initial data, it is possible to estimate the relative variability of each component spectrum separately. It is important because the corresponding solution spectra are distorted for the principal and low-intensity components to different degrees. Two procedures may be proposed for obtaining such estimates based on comparing the variants of solutions ("partial solutions") with the solution obtained for the original problem (hereafter, the reference solution) (Efron, 1979; Miller, 1974). Let us call **variants** either solutions of systems of linear equations obtained from the original one by discarding some conditional equations (**first procedure, or jackknife**) or parametrized solutions of the complete system (2.10) corresponding to different values of parameter α from relation (2.15) (**second procedure**). It is rather difficult to model all experimental errors because their distribution laws are often unknown, not to mention different systematic deviations.

In this case, estimating solution errors by modeling random errors in the initial data and comparing the perturbed solution-variants is not only inapplicable but also unreasonable. It allows one to count the stability of the solution of the already distorted problem. Here we consider forming data variants by discarding part of the original equations. The jackknife method allows us to vary all kinds of errors realized in the experiment: the noise of the spectrometer detector, instrument drift, violations of spectral additivity due to the interaction of components in mixtures, errors in determining the elements of $\tilde{C}, \widetilde{D}$, *etc*. Formation of systems of equations of less size than the initial one leads, naturally, to a specific increase of errors in the solution and a change in the condition number of the concentration matrix of the variant $\tilde{C}_B$. Experience in solving model problems has shown that condition $\mathrm{cond}(\tilde{C}_B) < 2\ \mathrm{cond}(\tilde{C})$ is acceptable in most cases. Parameterization (2.15), as already noted, does not improve individual solutions as compared to the LS, but in any case, when changing the value of α, the component spectra also change.

Model calculations have shown that the solution spectra, which change relatively little from variant to variant, are closer to the actual spectra of the components than those that vary considerably. All criteria discussed in Section 2.3 can serve as criteria for the spectral contour difference (or similarity).

To estimate the relative reliability of the individual spectra, they are ranked for each variant in ascending (or descending) order of the chosen similarity criterion. The average values of the ranks are calculated for the whole set of partial solutions. These average values are considered indicators of the relative variability of the spectra-solutions if the respective ranks are in good agreement with each other. The consent measure of the J ranks can be $J/2$ rank correlation coefficients for all possible pairwise rank combinations. The W_C (Johnson, 1977) concordance coefficient, a nonparametric statistical criterion of similarity, is better suited for our purposes.

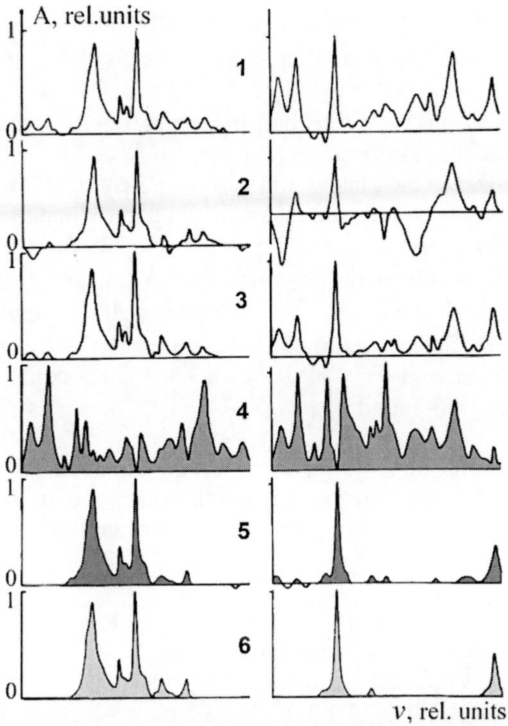

Figure 2.9. Decomposition of the IR spectra of 6-component mixtures with a priori knew estimate of relative concentrations obtained by gas chromatography. 1: the LS solution, 2:the regularized solution, 3: the weighted LS method, 4: the error spectra, 5: the corrected solutions, 6: the true spectra of substances. Left panel – isooctane (stable component, θ = 12.5°, the correlation coefficient r = 0.97), right side – benzene (unstable component, θ = 35.0°, r = 0.77).

Let the ranks for the J variants of the K-component mixture be obtained. The rank values are 1,2, ..., K. To calculate the W_C, all the J ranks are summed for each component spectrum, and then the mean value $\frac{J\cdot(K+1)}{2}$ is subtracted from each sum. Then, the squares of the resulting differences are summed.

The concordance coefficient W_C is calculated as follows

$$W_C = \frac{12\cdot(\textit{sum of squared differences})}{J^2\cdot K\cdot(K^2-1)}. \tag{2.19}$$

When the ranks coincide entirely, the value W_C = 1.0, and the more they differ, the W_C is closer to zero.

Assuming that the ranks are mutually independent, the value $H = \frac{(J-1) \cdot W_C}{1 - W_C}$ is distributed approximately as F_{n_1,n_2} with n_1=(*K*-1), n_2=(*K*-1)(*J*-1)-2 degrees of freedom. By comparing the values of *F* and *H*, we can estimate the significance (Johnson, 1977).

Table 2.2. shows the ranks for the 6-component problem **No. 2** (*Figure 1.4*). As is seen, the component ranks remain approximately the same from variant to variant. For this example $W_C = 0.78$; $H = 39.2$; $F_{n_1,n_2} = 2.4$, if using the value of the angle θ between vector-spectra calculated from (2.10), and $W_C = 0.83$; $H = 54.5$; $F_{n_1,n_2} = 4.8$. For these data, the significance of the concordance coefficient turned out to be extremely high, *i.e.*, the rankings are certainly valid, and one makes the conclusion about the "reliability" of the spectral contours found.

***Table 2.2.* Results of ranking the solution variants by the degree of increase in their variability relative to the reference solution. θ denotes the angle value between vectors as a difference criterion, *r* – correlation coefficient.**

Component number	Ranks for 12 solutions obtained for reduced data matrices	Mean rank	Resultant rank	Criterion of comparison
1	1 1 3 2 3 1 3 1 3 2 3 1	2.0	2	
2	5 5 6 5 6 6 6 5 6 5 6 6	5.8	6	
3	6 6 4 6 4 4 5 6 4 4 4 4	4.8	5	θ (1.39)
4	4 4 5 4 5 5 4 3 5 6 5 5	4.6	4	
5	3 3 2 3 2 3 2 4 1 1 1 2	2.3	3	
6	2 2 1 1 1 2 1 2 2 3 2 3	1.8	1	
1	1 1 2 2 2 2 2 1 3 2 3 1	1.8	2	
2	5 5 6 5 6 5 6 4 6 4 6 6	5.3	5-6	
3	6 6 5 6 5 6 5 6 5 5 5 4	5.3	5-6	*r* (1.40)
4	3 3 4 4 4 4 4 3 4 6 4 5	4.0	4	
5	4 4 3 3 3 3 3 5 2 3 2 3	3.2	3	
6	2 2 1 1 1 1 1 2 1 1 1 2	1.3	1	

However, stating that one spectrum is less reliable than the other is still insufficient for practice. More useful is detailed information about the "degree of fidelity" of individual bands (or segments) in the obtained spectra. The jackknife technique is the method that allows one to solve this problem. The method is based on the accumulation of absolute values of differences between the spectrum of the component of the reference solution (i.e., the solution obtained using the full set of data) and spectra from the partial solutions, normalized before subtraction by one of the peaks that are considered to be inherent in the spectrum of a given component. Usually, the most intensive band in the reference solution (if there are no other assumptions), which is maximally nonoverlapped with the bands of the other components, is taken as such a peak. To find the characteristic bands, one can analyze ratios of the spectral contours and select the spectral points by the maximum correlation between lines of the matrices $\widetilde{D}$ and $\widetilde{C}$ (for each component).

The accumulated absolute values of the differences can be considered a kind of "error spectra" for each component. There will be small amplitudes in places of characteristic bands in the error spectrum and outliers in areas of false bands. *Figure 1.20(4)* shows the error spectra for the model 6-component problem we are considering as an example.

The estimates of the relative reliability of partial regions of the component spectra can be used to correct spectral contours by the target transformation technique considered above. The process is based on the assumption that the errors of spectroscopic measurements are minor. Then, the mixture spectra can be considered sufficiently accurate linear combinations of the component spectra. *Figure 2.9(5)* shows the result of such correction for two of the six components in the mixture.

The correction procedure consists in removing spectral regions corresponding to large relative amplitudes in the error spectra (such regions are assigned zero intensities) followed by projection of the corrected vectors-spectra $\boldsymbol{t}$ onto the mixture matrix using the target transformation considered in Section 2.3 (equation 2.1). The resulting projection $\boldsymbol{p}$ is a linear combination of the mixture spectra vectors and thus may be considered a component contour. The projection spectrum is as close as possible to the original vector (in the Euclidean metric), i.e., the minimal perturbation of the spectrum-solution is guaranteed for this transformation. The latter, in turn, means that all the corrections made will be taken into account as fully as possible. The exclusion of false segments can be checked by the angle value or the correlation coefficient between the $\boldsymbol{t}$ and $\boldsymbol{p}$ vectors. If the projection significantly differs from the tested spectrum (in practice, one can usually take $\theta > 15^{o}$ or $r < 0.95$ as a boundary), it means that among the excluded bands in the obtained spectrum, there were bands inherent to the given component. Such bands can be recognized from a visual comparison of the $\boldsymbol{t}$ and $\boldsymbol{p}$ contours since the excluded "right" bands reappear in the projection spectrum. In addition, a significant difference between $\boldsymbol{t}$ and $\boldsymbol{p}$ may indicate the incomplete elimination of false segments.

The discrepancy between the corrected and actual spectra of the components is directly related to the Alentsev-Fok conditions of the spectral nonoverlap. If these conditions are strictly satisfied and there are no errors in the matrix, it is possible to find spectra of the components that coincide with the true ones.

Of great interest is when the nonoverlap conditions are not satisfied. Then, when using the considered tactics of eliminating false bands, it is possible to find more "simple" than in reality spectra of the components. However, in most cases, the individual spectra contain areas with bands belonging to different components that do not coincide in the position of their maxima. In this case, the tactic of excluding the false band consists in smoothing the resulting spectral contour in the vicinity of its maximum.

Additional criteria for solution correctness are the non-negativity of the elements of the corresponding concentration matrix and the correspondence of the component concentration curves to the separation method. The difference matrix $D^{\#}$ elements should not exceed the experimental error level in the matrix $\widetilde{D}$. The target transformation (2.1) allows one to carry out the input control of the initial information if the experimental errors are not too large.

Summarizing the material of this section, one can say the following.

1. The problem of searching for component spectra from sets of additive spectra of related mixtures is mathematically ill-posed and, in general, ambiguous if nonoverlap conditions are not met.
2. Even in the case of the valid nonoverlap conditions for individual spectra, the solution depends on the choice of values of the target function parameters.
3. In these conditions, the usual statistical estimations of the solution variance, based on the initial data variance estimates, are inapplicable since the errors also include the unknown a priori systematic deviations that violate the additivity of data.
4. It is recommended to evaluate the reliability of the solution by a comparative analysis of the contours of the component spectra obtained by varying:
   - starting approximation,
   - target function parameters (both set and values),
   - the completeness of the set of initial data (*e.g.*, by obtaining solutions from incomplete sets of spectra through all combinations).

The algorithms for the analysis of the additive mixture spectra may be easily adapted to more general cases and applied to the study of nonnegative multi-component data of different nature.

References

Dennis, J. E. Jr., Gay D. M., Welsh, R. E. (1981). *ACM Trans. on. Math. Soft.*, **7,** No. 3, 348–368.

Dennis, J. E. Jr., Gay D.M., Welsh, R. E. (1981). *ACM Trans. on. Math. Soft.*, **7,** No. 3, 369–383.

Dennis, J. E. Jr., Schnabel, R. B. (1983). *Numerical Methods for Unconstrained Optimization and Nonlinear Equations,* New Jersey: Prentice-Hall.

Efron, B. (1979). *The Annals of Statistics,* **7**, No. 1, 1-26

Fok, M. V. (1973). Separation of Complex Spectra into Individual Bands by the Generalized Alentsev Method. *In: The Lebedev Physics Institute Series. Luminescence and Nonlinear Optics. Ed. by D. V. Skobel'tsyn.* Springer-Verlag US, Boston, MA. 286p. (1-22).

Gill, P. E, Murray, W., Wright, V. H. (1981). *Practical Optimization.* Academic Press, San Diego, CF 92101.

Johnson, N. L., Leone, F. C. (1977). *Statistics and Experimental Design in Engineering and the Physical Sciences. V. 2 (Probability & Mathematical Statistics)*, New York, John Wiley & Sons.

Knorr, F. J., Harris, J. M. (1981). *Anal. Chem.*, **53**, No. 2, 272–276.

Lawson, Ch. L., Hanson, R. J. (1974). *Solving Least Squares Problems*, New Jersey: Prentice–Hall, Inc., Englewood Cliffs.

Lawton, W. H., Sylwestre, E. A. (1971). *Technometrics*, **13**, No. 3, 617–633.

Meister, A. (1984). *Anal. Chim. Acta*, **161**, 149–161.

Miller, R. G. (1974). *Biometrica*, **61**, No. 1, 1-15.

Strang, G. (1980). *Linear Algebra and its Applications*, Academic Press.

Sylvestre, E. A., Lawton, W. H., Maggio, M. S. (1974). *Technometrix*, **16**, No. 3, 353 – 368.

Voevodin, V. V. (1983). *Linear Algebra*, Mir Publishers.

Volkov, V. V. (1996). *Applied Spectroscopy*. **50,** No. 3, 320 – 326.

CHAPTER THREE

THE X-RAY DIFFRACTION TOMOGRAPHY

3.1. Introduction

The high-resolution X-ray diffraction technique (HRDT) has been intensively developed in the past decades to diagnose the new-fabricated nanoscale semiconductor materials. In particular, such localized nanoscale clusters are the quantum dots and or wires, and hence, they are related to novel electronic and optical properties of semiconductors. A number of the experimental and theoretical works (Pavlov & Punegov, 2000; Benediktovitch, Feranchuk & Ulyanenkov, 2013; Seregin *et al.*, 2019) using the HRDT have been done, where the statistical dynamical theory of the X-ray diffraction scattering in crystals has been exploited as a tool for investigating.

In the present, the X-ray diffraction tomography (XRDT) gets a special nomination to investigate the point crystal defects due to the high sensitivity of the XRDT method to relatively small alternations of the crystal lattice parameter caused by various crystal defects like striations, grain boundaries, stacking faults, clusters, and dislocations (Authier, 2001). All such defects change the positions of individual atoms in the crystal lattice cell relative to their usual places. In practice, the Laue and Bragg diffraction techniques are used for investigating the actual crystal structures that yield the 2D projections of a crystal. Usually, to interpret and analyze such the experimental 2D images, the latter is needed for comparing with the calculated 2D ones using the Takagi–Taupin equations (Takagi, 1962; Taupin, 1964; Epelboin, 1975; Besedin, Chukhovskii & Asadchikov, 2014).

Historically, the ab initio XRDT technique originates from the X-ray diffraction topography (Guinier, 1956; Authier, Lang, 1964; Tanner, Phil 1966; Bowen, Tanner, 1998). Afterward, Ludwig *et al.* (2001) have first shed a bright light on the capability of the *ab initio* XRDT technique. Danilewsky *et al.* (2011) and Hänschke *et al.* (2012, 2017) have exhibited the synchrotron radiation capacity to obtain a look at the 3D images dislocation loops in crystalline wafers. Asadchikov *et al.* (2018) have used the XRDT technique to identify sole dislocations distributed within a single crystal Si.

According to the XRD technique, one rotates the crystal sample around an axis perpendicular to a set of reflection crystal planes; In other words, the sample rotation axis is chosen to coincide with the diffraction vector **h**. In turn, it means that one can obtain the 2D X-ray diffraction projections for different sample rotation angles Φ. Such 2D Φ-projections correspond to a particular orientation of the X-ray diffraction planes relative to the intrinsic crystal Cartesian system of coordinates (see *Figure 3.1*).

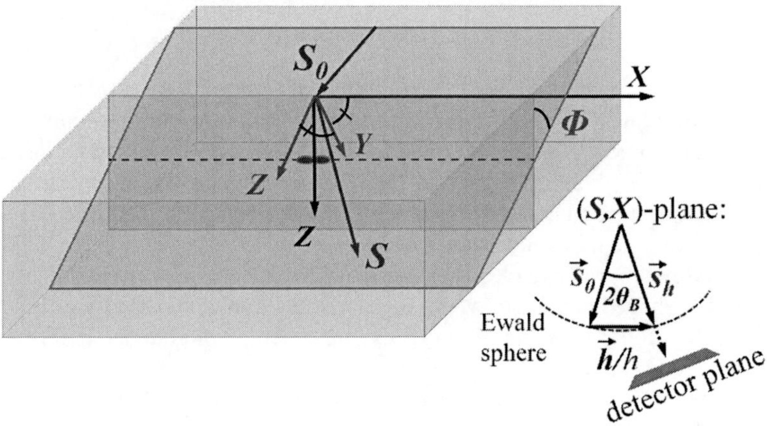

Figure 3.1. Schematic X-ray diffraction scattering by the crystal point defect in the Φ-inclined diffraction plane. Sign ••• denotes the point defect location. Φ is the sample rotation angle around diffraction vector $\boldsymbol{h} \parallel$ axis $0X$. The symbols S_0, S_h, and X are the oblique system coordinates in the X-ray diffraction plane. The axes $0Z$ and $0Y$ are perpendicular to axis $0X$, respectively. The unit vector $\boldsymbol{s}_0$ is directed to the wave vector $\boldsymbol{k}_0$ of the incident monochromatic plane wave.

Thus, the *ab initio* XRDT technique collects a set of diverse 2D image patterns (2D-IPs) that relate to different rotation angles Φ of the crystal samples around diffraction vector $\boldsymbol{h}$. The next point of the *inverse* XRDT problem solution is the subsequent recovery of the 3D displacement field function $f(\mathbf{r}) = \boldsymbol{h}\cdot\mathbf{u}(\mathbf{r})$ ($\mathbf{u}(\mathbf{r})$ is the displacement vector due to the crystal defect).

The first XRDT experiments have been performed on synchrotron radiation sources ESRF (Ludwig *et al.*, 2001) and SPring-8 (Kawado *et al.*, 2004). Similarly, such experiments were carried out on laboratory setups using the X-ray characteristic $K_{\alpha1}$-line radiation (Asadchikov *et al.*, 2018).

It is worth noticing that because the 2D-IP image contrast relates to different regions around defects, the 3D recovery problem of the displacement field function $f(\mathbf{r})$ relates equally, if not to a greater extent, to decoding the 2D-IPs data close to the region of defect core. Evidently, in turn, processing *ab initio* 2D-IPs data requires the use of a sound theoretical basis to plausible recovering the 3D displacement field function $f(\mathbf{r})$. In parallel, it is essential to build such the analytical solutions of the Takagi–Taupin equations, which would make it possible to describe the image features near around a single defect with sufficient accuracy.

Below, we exhibit a capability to solve the inverse XRDT problem for obtaining mesoscopic scale function $f(\mathbf{r})$ on an example of the Coulomb-types point defect located at the point $\mathbf{r}_0$ inside a crystal. Choosing such the crystal lattice point defect is of definite interest since these point defects (e.g. interstitial-types clusters) affect the electronic properties of semiconductor materials. Hence, here, the XRDT technique could powerfully be applied.

Following the theory of isotropic elastic media, for the Coulomb-types point defect located at the point $\mathbf{r}_0$ within a crystal, the 3D-defined displacement field function $f_{Ctpd}(\mathbf{r}-\mathbf{r}_0)$ has the form

$$f_{CPD}(\mathbf{r}-\mathbf{r}_0) = \frac{F}{4\pi}\frac{\boldsymbol{h}\cdot(\mathbf{r}-\mathbf{r}_0)}{|\mathbf{r}-\mathbf{r}_0|^3+\varepsilon}, F = \text{const}, \varepsilon \rightarrow +0. \tag{3.1}$$

We show that sometimes the simulated annealing (SA) (Kirkpatrick *et al.*, 1983) and quasi-Newton gradient descent (qNGD) algorithms (Gill *et al.*, 1981; Dennis *et al.*, 1981) work to recover the 3D-defined displacement field function $f_{CPD}(\mathbf{r}-\mathbf{r}_0)$ Konarev, Chukhovskii & Volkov, 2019). The Coulomb-types point defect (CPD) is located in the middle of Si(111) crystal, the diffraction vector $\mathbf{h}$ = $[\bar{2}20]$, the incident X-ray radiation is characteristic $MoK_{\alpha 1}$, λ=0.071 nm. The CPD can be treated as a pure phase object in our case. As to the theoretical framework, we use the semi-kinematical approach for building out an analytical solution of the Takagi-Taupin equations (Takagi, 1962, 1969; Taupin, 1964).

Several computer iterative algorithms could be tested to solve the *inverse* XRDT problem, such as the quasi-Newton-Levenberg-Marquardt and Simulated Annealing (Moore, 1978; Kirkpatrick, Gelatt & Vecci, 1983, respectively).

In general, to recover the function f_{CPD} ($\mathbf{r}$-$\mathbf{r}_0$), its numerical values might be assumed as a multidimensional vector $\mathcal{P}$ *in search*.

In principle, the observed 2D-IPs can be simulated using numerical solutions of the Takagi-Taupin equations for any function $f(\mathbf{r})$. Unfortunately, such an approach requires cumbersome numerical calculations and takes much CPU time for treating the inverse XRDT problem. Nevertheless, a door to go along this way is open, and this is a good topic for future work. In general, to recover the function f_{CPD} ($\mathbf{r}$-$\mathbf{r}_0$) *in search,* its numerical values might be considered a multidimensional vector $\mathcal{P}$ *in search.*

On the other hand, such a choice of the vector $\mathcal{P}$ is meaningful only if some hard constraints are imposed. The significant restrictions are equivalent to choosing the function f_{CPD} ($\mathbf{r}$-$\mathbf{r}_0$) *in search* in the analytical form (3.1) and hence, to selecting the vector $\mathcal{P}$ *in search* as $\mathcal{P} = \{p, \mathbf{r}_0, F\}$, where parameter p characterizes the decreasing power of the function $f_{CPD}(\mathbf{r}\text{-}\mathbf{r}_0)$ located at the point $\mathbf{r}_0$ and parameter F is the defect power. As to the location vector $\mathbf{r}_0$, only the coordinate z_0 takes a sense. Indeed, one can readily determine the lateral coordinates (x_0, y_0) at the XRDT 2D-IPs.

In the beginning, we are confined by the case of the rotation (inclination) angle $\Phi = 0$, and without the loss of generality, one may put on $x_0 = y_0 = 0$.

Hence, in the case of the CPD function (3.1), the *inverse* XRDT problem reduces to retrieving three-component vector $\mathcal{P} = \{p, z_0, F\}$ in search using the sole 2D-IP data.

To solve the *inverse* XRDT problem under consideration, we apply the Simulated Annealing (SA), and quasi-Newton Gradient Descend (qNGD) iterative algorithms.

In the further, in the case of several 2D-IPs with including the rotation angles $\Phi \neq 0$ we analyze and test the combined quasi-Newton-Levenberg-Marquardt–Simulated Annealing (qNLMSA) algorithm (Chukhovskii, Konarev & Volkov, 2020) to optimize the XRDT target function in a χ^2-sense. Also, we will demonstrate that using the (qNLMSA) algorithm, the absolute (global) minimum of the XRDT target function, $\chi^2 = 0$, may be achieved in the case of the noise-free 2D-IPs. The latter is a persuasive argument to use the XRDT technique with the low counting statistics 2D-IPs to obtain digital information about nanoscale clusters in crystals.

At last, based on the corresponding Gaussian noisy 2D-IPs data, we will bring and discuss preliminary results for recovering the 3D function $f_{CPD}(\mathbf{r} - \mathbf{r}_0)$.

3.2. Theoretical background

In the case of symmetric Laue diffraction, the Takagi-Taupin (TT) equations can be written in the form (Takagi, 1962, 1969; Taupin, 1964)

$$\begin{cases} -i\dfrac{\lambda}{\pi}\dfrac{\partial E_0}{\partial s_0} = \chi_0 E_0 + C\chi_{\bar{h}} e^{if(\mathbf{r})} E_h - i\dfrac{\lambda}{\pi}\delta(s_0 + s_h) e^{i\frac{\pi}{2\lambda}\alpha(s_h - s_0)} \\ -i\dfrac{\lambda}{\pi}\dfrac{\partial E_h}{\partial s_h} = \chi_0 E_h + C\chi_h e^{-if(\mathbf{r})} E_0 \end{cases} \tag{3.2}$$

Here, $E_0 \equiv E_0(s_0, s_h)$ and $E_h \equiv E_h(s_0, s_h)$ are transmitted and diffracted wave amplitudes of the electric wave field propagating within an imperfect crystal. Equations (3.2) describe the two-beam X-ray Laue diffraction within the scattering plane (s_0, s_h). Accordingly, the coefficients χ_0, χ_h and $\chi_{\bar{h}}$ are the zero, $\boldsymbol{h}$,&$-\boldsymbol{h}$ – Fourier components of the perfect crystal electric susceptibility. The function $f(\mathbf{r})$ is the 3D-defined defect displacement field. C is the X-ray polarization factor equal to unity for σ-polarization and $\cos 2\theta_B$ for π-polarization (remind that θ_B is the Bragg angle); λ is the X-ray wavelength. As is seen, the 3D-defined function $f(\mathbf{r})$ inserts into the TT equations (3.2) *via* the exponential phase factors of the coefficients χ_h and $\chi_{\bar{h}}$. Parameter α characterizes the deviation of the incident plane wave from the exact Bragg condition, $\alpha = (2k_0\mathrm{h}+h^2)/2k^2$; k_0 is the incident X-ray wave vector, $k = 2\pi/\lambda$.

Inclusion of the δ-function term into the right-hand side of the first equation of (3.2) ensures that amplitudes $E_0(s_0, s_h)$ and $E_h(s_0, s_h)$ satisfy the boundary conditions on the entrance crystal surface, $s_0 = -s_h$,

$$E_0(s_0, s_h)\Big|_{s_0 = -s_h} = e^{i\frac{\pi}{\lambda}\alpha s_h}, \quad E_h(s_0, s_h)\Big|_{s_0 = -s_h} = 0 \qquad (3.2a)$$

Substituting of $E_0(s_0, s_h)$ and $E_h(s_0, s_h)$ as follows

$$\begin{aligned} E_0(s_0, s_h) &\to E_0(s_0, s_h)e^{i\chi_0\frac{\pi}{\lambda}\left(s_0+s_h\right)}, \\ E_h(s_0, s_h) &\to E_h(s_0, s_h)e^{i\chi_0\frac{\pi}{\lambda}\left(s_0+s_h\right)} \end{aligned} \qquad (3.3)$$

and introducing the dimensionless coordinates, one reduces equations (3.2) to the form

$$\begin{cases} \dfrac{\partial E_0}{\partial S_0} = i\Gamma_{\bar{h}}e^{if(\boldsymbol{r})}E_h + \delta(S_0+S_h)e^{iw(S_h - S_0)} \\ \dfrac{\partial E_h}{\partial S_h} = i\Gamma_h e^{-if(\boldsymbol{r})}E_0 \end{cases} \qquad (3.4)$$

Aiming at the inverse XRDT problem solution, we exploit a semi-kinematical solution of the TT equations (3.4) (Chukhovskii, Konarev & Volkov, 2019; 2020)

$$E_h(S_0, S_h) = -i(1-i\kappa)\int_{-S_0}^{S_h} dP_h e^{-if(S_0, P_h)}E_0^{(idea)}(S_0, P_h), \qquad (3.5)$$

where the transmitted and diffracted wave amplitudes $E_0^{(\text{idea})}(S_0, S_h)$ and $E_h^{(\text{idea})}(S_0, S_h)$ of the wavefield propagating within a perfect crystal are defined by

$$E_0^{(idea)}(S_0,S_h)=e^{iw(S_h-S_0)}\left\{\begin{array}{l}\dfrac{iw\sin\left[(S_0+S_h)\sqrt{(1-i\kappa)^2+w^2}\right]}{\sqrt{(1-i\kappa)^2+w^2}}+\\ \cos\left[(S_0+S_h)\sqrt{(1-i\kappa)^2+w^2}\right]\end{array}\right\}, \quad (3.6a)$$

$$E_h^{(idea)}(S_0,S_h)=\frac{-i(1-i\kappa)}{\sqrt{(1-i\kappa)^2+w^2}}e^{iw(S_h-S_0)}\sin\left[(S_0+S_h)\sqrt{(1-i\kappa)^2+w^2}\right]. \quad (3.6b)$$

where the following notations are introduced

$$S_0=\frac{s_0}{\ell}, S_h=\frac{s_h}{\ell}, \Gamma_h=\Gamma_{-h}=-(1-i\kappa),$$
$$\ell=\frac{\lambda}{\pi[Re(C^2\chi_h\chi_{-h})]^{\frac{1}{2}}}, w=\frac{\alpha}{[Re(C^2\chi_h\chi_{-h})]^{\frac{1}{2}}}, \quad (3.6c)$$
$$s_0+s_h=\frac{Z}{\cos\theta_B}, s_h-s_0=\frac{X}{\sin\theta_B}, 0<\kappa\ll 1.$$

Before proceeding further, we clarify the formulation of the inverse XRDT problem keeping in mind developing the iterative computer procedure for optimizing the XRDT target χ^2-function.

As was pointed out by Authier (2001), in some parts of the XRDT 2D-IPs, contrast is due to the X-ray wavefield propagating near around the defect core, *i.e.,* the strongly distorted crystal lattice region. It does mean that the direct defect contrast on the XRDT 2D-IPs arises due to the X-ray wavefield scattering nearby the defect core. This circumstance allows treating the *inverse* XRDT problem in the context of direct defect contrast on the XRDT 2D-IPs described by equation (3.5).

Thus, equation (3.5) describes the *direct* defect image contrast at the 2D-IPs in terms up to the first-order perturbation theory solution of the Takagi-Taupin equations (3.2) inclusively. In the form of (3.6a) and (3.6b), the wavefield amplitudes are the zero-order perturbation theory solutions of the TT equations for a perfect crystal.

The semi-kinematical solution (3.5) describes the X-ray wavefield scattering nearby the defect core linked with the direct defect image at the 2D-IPs that is why we claim both the answer (3.5) and the expressions (3.6a) and (3.6b) as a theoretical framework for treating the *inverse* XRDT problem in question. Accordingly, since the 3D $f_{\text{CPD}}(\mathbf{r}-\mathbf{r}_0)$-function goes to zero far away from the defect core, when $|\mathbf{r} - \mathbf{r}_0| \to \infty$, solution (3.5) tends to expression (3.6b) for the diffracted amplitude $E_h^{(idea)}(S_0, S_h)$.

Ultimately, based on the optimizing the XRDT target χ^2-function, the *inverse* XRDT problem is formulated as follows, namely: to retrieve the three-component vector $\boldsymbol{\mathcal{P}}^{(true)} = \{p, z_0, F\}$ of the 3D-defined function $f_{\text{CPD}}(\mathbf{r} - \mathbf{r}_0)$ *in search* by using the artificial 2D-IPs data $I_{h,obs}[\text{x(T)}, \text{y(T)}; \Phi_i]$. In turn, they are simulated for various rotation angles Φ_i, i =1, 2, 3, and N (N is the total number of the 2D-IPs) using the semi-kinematical solution (3.5).

For convenience, we rewrite the solution (3.5) in the explicit form using the dimensionless Cartesian coordinates X, Y, Z through their values at an exit crystal surface.

Then, the equation (3.5) takes the form

$$E_h(\text{x(T)}, \text{y(T)}) = -\frac{i(1 - i\kappa)}{\cos\theta_B \cos\Phi} \int_0^{\text{T}} dz e^{-if_{CPD}(\text{x(z)}, \text{y(z)}, \text{z})} E_0^{(idea)}\left(\frac{\text{x(z)}}{\sin\theta_B}, \frac{\text{z}}{\cos\theta_B \cos\Phi}\right) \qquad (3.7)$$

where the following notations are introduced

$$\text{x(z)} = \text{x(T)} - (\text{T} - \text{z})\frac{tg\,\theta_B}{\cos\,\Phi}, \; \text{y(z)} = \text{y(T)} - (\text{T} - \text{z})tg\,\Phi. \qquad (3.8)$$

Accordingly, the formula (3.1) for the 3D-defined CPD function takes the form

$$f_{\text{CPD}}(\text{x(z)}, \text{y(z)}, \text{z}) = G\frac{\text{x(z)}}{\left(\text{x}^2(\text{z}) + \text{y}^2(\text{z}) + (\text{z - z}_0)^2\right)^{3/2} + \varepsilon}, \quad \varepsilon \to +0 \cdot \qquad (3.9)$$

In the next Section, the following numerical values are used in calculating the XRDT data $I_{h,obs}[\text{x(T)}, \text{y(T)}; \Phi_i]$, namely: we assume that diffraction vector $\boldsymbol{h}$ = $[2\bar{2}0]$, the X-ray wavelength $\lambda = 0.0709$ nm, the Bragg angle θ_B=10,65°. The susceptibility coefficients $Re(\chi_h) = Re(\chi_{-h}) = -1.921 \times 10^{-6}$ and $\text{Im}(\chi_h) = \text{Im}(\chi_{-h}) = 1.55 \times 10^{-8}$; the extinction length, $\Lambda = \pi\ell\cos\theta_B$, Λ =36.287 μm. The CPD is incorporated within Si(111)-crystal on the depth z_0=T/2, T is the dimensionless sample thickness in units ℓ. The CPD power parameter F is equal to 0.064 μm³.

Then, in the formula (3.9), the constant $G = \left(\frac{\sin\theta_B F}{\lambda\ell^2}\right)$ takes the value of the order of unity.

Thus, for the given set $\{\Phi_i\}$, the artificial XRDT 2D-IPs $\{I_{h,obs}[\text{x(T)}, \text{y}(T); \Phi_i]\}$ are evaluated. The latter are used in the process of optimizing the XRDT target function $\mathcal{F}\{\boldsymbol{\mathcal{P}}\}$ in a χ^2–sense, *i.e.*

$$\chi^2 = \frac{1}{n\{\mathrm{x(T)}*\mathrm{y(T)}\}} \sum_{i=1}^{n} n(\Phi_i) \cdot$$
$$\sum_{\{\mathrm{x(T),y(T)}\}} \frac{\left(I_{h,obs}[\mathrm{x(T),y(T)};\Phi_i] - I_{h,calc}[\mathrm{x(T),y(T)};\Phi_i]\right)^2}{I_{h,obs}[\mathrm{x(T),y(T)};\Phi_i]^2} = min, \qquad (3.10)$$
$$n(\Phi_i) = 1, i = 1, \ldots n,$$

where the rectangular sample prism with dimensions $0 \leq z \leq T$, $-x/2 \leq x(T) \leq x/2$, $-y/2 \leq y(T) \leq y/2$, the sum in the right-hand side of equation (3.10) is taken over an array of the XRDT 2D-IPs, *n* is the total number of the inclined XRDT 2D-IPs.

In the subsequent Section, one applies and discusses the implementation of the SA, qNGD, and the combined qNLMSA iterative algorithms for solving the *inverse* XRDT problem in question.

3.3. The computer iterative algorithms

As is known, the Simulated Annealing (SA) algorithm (Kirkpatrick *et al.*, 1983) is widely used to minimize nonlinear objective χ^2-functions and applied to problems with a large number of variables and combinatorial nature of iterative calculations.

We will distinguish the two system states: the *current* and *trial* values of the 3D function $f_{CPD}(\mathbf{r} - \mathbf{r}_0)$. When the *current* state becomes a *trial* one in the iterative process, it is used further as a new *system current state*.

A pseudorandom change of system is applied to the *current* state, becoming a *trial* function $f_{CPD}(\mathbf{r} - \mathbf{r}_0)$. The probability of accepting the *trial* function as a new *current* one is determined by the value of the Boltzmann function $\exp(-\Delta\chi^2/\tau)$, where $\Delta\chi^2$ is the change in the χ^2-function and τ is the temperature parameter. The choice of a new *current* system state depends strongly on the τ-value. As is easily seen, a higher temperature means a higher value of probability $\exp(-\Delta\chi^2/\tau)$); due to which the *trial* mode can be accepted as the new *current* one even in the case of worse *trial* mode as compared with the *current* system state, when $\Delta\chi^2 > 0$. However, it $\Delta\chi^2 < 0$, the *trial* mode is taken as a new *current* one. When starting the iterative minimization of the χ^2-function, temperature τ is chosen sufficiently high, so that frequency of accepting a worse state of the system as a *current* one exceeds the state improvement rate by a factor of 10–100 (in other words, the changes of the system are almost random). This circumstance makes the SA program "roam" in space, searching for minimum values of the system mode $f_{CPD}(\mathbf{r} - \mathbf{r}_0)$. In general, the SA program may overcome local minima of the target χ^2-function, which are one of the main hindrances for many other nonlinear methods used to minimize the target χ^2-function, *e.g.*, for gradient descent methods (Gill *et al.*, 1981; Dennis *et al.*, 1981).

To minimize the target χ^2-function, the temperature τ_{n+1} at the step after the previous nth step (external cycle over n) decreases monotonically as $\tau_{n+1} = F\cdot\tau_n$, where the annealing coefficient F is equal to 0.9–0.95.

Table 3.1. **The SA iterative algorithm code protocol**

Protocol SA stages	Functional SA stages
initial number of elements of the system $\{i, j, k\}$ temperature $\tau_1 = \tau_{in}$ $\chi_{1,1}^2 = \chi_{in}^2$	$f_{1,1}(\{i, j, k\}) = f_{in}(\{i, j, k\})$ n is the cycle number with temperature τ_n; m is the iteration number in the nth cycle; $m = (1, 2,.., M)$, $M = 500000$ is the total iteration number, $\tau_{n+1} = F\tau_n$, $F = 0.95$.
cycle number $n = \{1, 2,.., N\}$, $N = 200$	$f_{m+1,n} = f_{m,n}(\{i, j, k\}) \mid U\{i, j, k\} \rightarrow \hat{U}\{i, j, k\}$ for one element $U\{i, j, k\}$; energy change $\Delta_{m,n} = \chi_{m+1,n}^2(f_{m+1,n}) - \chi_{m,n}^2(f_{m,n})$
m is the iteration number in the nth-cycle, $m = (1, 2,.., M)$, M=500000	$\Delta_{m,n} < 0$: the *trial* state $f_{m+1,n}$ is taken undoubtedly; $\Delta_{m,n} > 0$: the *trial* state $f_{m+1,n}$ is taken, if the probability $W_{m,n} = \exp(-\Delta_{m,n} / \tau_n)$ exceeds a random generated number in the unit range [0,1]. In opposite, one takes the state $f_{m,n}$.
if $N = 200$, the SA code stops	If frequency ν of the prosperous state $f_{m,n}$ changes for τ_n is larger than $M/10$, the SA code passes to the next step in temperature τ_{n+1}. If $\nu < 50$ for τ_n or $n = N$, the SA stops fixing the state $f_{m,n}$.

The value of the target χ^2-function decreases similarly to the decrease in the system internal energy upon cooling (the so-called Boltzmann machine, Metropolis *et al.*, (1953)).

According to work (Kirkpatrick, 1983, Metropolis *et al.*, 1953), some details of the SA iterative SA algorithm code protocol are collected in *Table 3.1*.

The essential advantage of the SA iterative algorithm is its capacity to go out the target χ^2–function from the local minima of the target χ^2–functions. In turn, the latter is some imperfection for many nonlinear optimization codes, particularly for the quasi-Newton Gradient Descend (qNGD) ones (Gill *et al.*, 1981; Dennis *et al.*, 1981).

First, we test the SA algorithm code to be effective in the case of the single XRDT 2D-IP when the sample rotation angle Φ =0.

Correspondingly, formula (3.10) is simplified to the form

$$\chi^2 N\{\mathrm{x(T)},\mathrm{y(T)}\}\cdot$$
$$\cdot\sum_{\{\mathrm{x(T)},\mathrm{y(T)}\}}\frac{\left(I_{h,obs}[\mathrm{x(T)},\mathrm{y(T)};\Phi=0]-I_{h,calc}[\mathrm{x(T)},\mathrm{y(T)};\Phi=0]\right)^2}{I_{h,obs}[\mathrm{x(T)},\mathrm{y(T)};\Phi=0]^2}=\mathrm{Min},$$
$$I_{h,calc}\{\mathrm{x(T)},\mathrm{y(T)};\Phi=0\}=|E_h\{\mathrm{x(T)},\mathrm{y(T)};\Phi=0\}|^2, \tag{3.11}$$

where $I_{h,obs}\{\mathrm{x(T)},\mathrm{y(T)};\Phi=0\}$ is the true X-ray topography 2D-IP simulated using formulae (3.7)-(3.9). The 2D-IP $I_{h,calc}\left\{\begin{matrix}\mathrm{x(T)},\mathrm{y(T)};\\ \Phi=0\end{matrix}\right\}$ should be subsequently evaluated using the *trial* system modes $f_{\mathrm{CPD}}(\mathbf{r}-\mathbf{r}_0)$ in the target $\chi 2$–function optimization fashion. Without losing the meaning of the overall results, we put the scaling coefficient G=1 in formula (3.9).

Besides, to estimate the degree of the iterative process convergence, we will use the 'error' parameter CP, which is defined as

$$CP=\frac{1}{\{N\{\mathrm{x,y,z}\}}\sum\nolimits_{\{\mathrm{x,y,z}\}}\frac{|f_{obs}\{\mathrm{x,y,z}\}-f_{calc}\{\mathrm{x,y,z}\}|}{|f_{obs}\{\mathrm{x,y,z}\}|} \tag{3.12}$$

It yields the averaged evaluation of the deviation of a *current* mode $f_{\mathrm{CPD}}(\mathbf{r}-\mathbf{r}_0)$ from the *true* solution $f_{\mathrm{obs}}(\mathbf{r}\text{-}\mathbf{r}_0)$.

The true XRDT 2D-IP is shown in *Figure 3.2*.

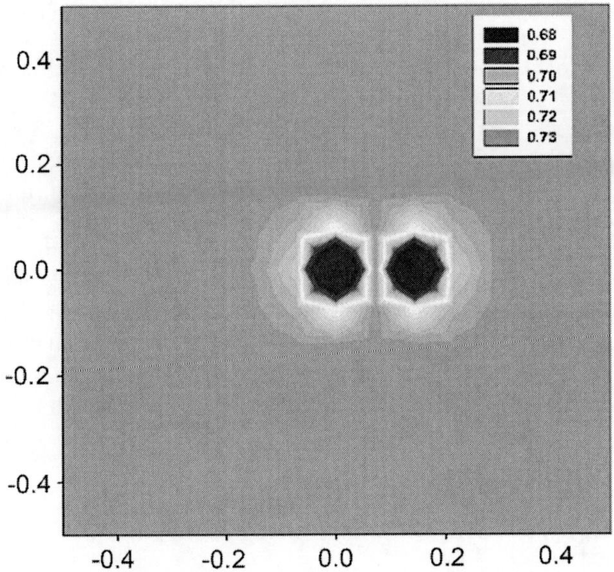

Figure 3.2. The observed XRDT 2D-IP: the Coulomb-types point defect in the crystal sample Si(111), the true vector $\boldsymbol{\mathcal{P}}$ = {1.5, 0.5T, 1}. The computer simulation has been carried out over the grid brick nodes {15, 15, 15}, $-0.5 \leq x \leq 0.5$, $-0.5 \leq y \leq 0.5$, T=1.

In *Table 3.2*, based on the SA iterative algorithm code for different spatial grids, the calculated results of recovering the 3D function $f_{\mathrm{CPD}}(\mathbf{r}-\mathbf{r}_0)$ are listed.

***Table 3.2.* Initial and ultimate data for the 3D function $f_{\mathrm{CPD}}(\mathbf{r}-\mathbf{r}_0)$. The initial analytical functions $f_{\mathrm{CPD},i}(\mathbf{r}-\mathbf{r}_0)$ are given by parameters p_i, $i = 1 \div 4$, G=1, z_0=0.5.**

spatial grids $\{i, j, k\}$	initial state			Ultimate state		
	p_i, $\{i = 1 \div 4\}$	χ^2	CP	p_i, $\{i = 1 \div 4\}$	χ^2	CP
{41,41,41}	{0.9}	0.79	1	{1.5}	$8 \cdot 10^{-7}$	$4 \cdot 10^{-7}$
{41,41,41}	{0.5,1.0,1.8}	0.48	0.96	{1.49,1.49,1.52}	$3 \cdot 10^{-4}$	$1 \cdot 10^{-2}$
{41,41,41}	{0.9,1.2,1.8,2.1}	1.44	1	{1.48,1.54,1.49,1.49}	$3 \cdot 10^{-3}$	$1.4 \cdot 10^{-2}$

As some constraints, we only will use the symmetry properties of the $f_{\text{CPD}}(\mathbf{r}-\mathbf{r}_0)$- function over the variables (x, y), and (z – z_0), and the requirement of a monotonic decrease with increasing of variables $|x|$, $|x|$, and $|z - z_0|$ is only imposed.

The artificial XRDT 2D-IP for the crystal Si(111) in Fig. 3.2 has been calculated on the spatial grid nodes $\{15 \times 15 \times 15\}$. As is seen, the observed XRDT 2D –IP is symmetric in regards to coordinate y and shifts from the 2D –IP center along coordinate x as a whole by $0.5T{\cdot}\tan\theta_B$. The initial data and the recovery results of the 3D $f_{\text{CPD}}(\mathbf{r}-\mathbf{r}_0)$-functions are listed in *Table 3.3.* There are the spatial grids in the first column, on which calculations have been carried out. The *bold* figure levels along the z-axis refer to the varying numerical values of the 3D-defined function $f_{\text{CPD}}(\mathbf{r}-\mathbf{r}_0)$. For other levels pointed out, the numerical values of the 3D-defined function $f_{\text{CPD}}(\mathbf{r}-\mathbf{r}_0)$ *in search* are fixed and coincide with ones of the 3D-defined function $f_{\text{CPD,true}}(\mathbf{r}-\mathbf{r}_0)$.

Table 3.3. **Initial and ultimate of the function $f_{\text{CPD}}(\mathbf{r}-\mathbf{r}_0)$. The ultimate $f_{\text{CPD}}(\mathbf{r}-\mathbf{r}_0)$ has been obtained using both the qNGD and SA algorithms.**

spatial grid $\{i, j, k\}$	initial			ultimate	
	p_{ini}	target χ^2-function	CP	target χ^2-function	CP
{15, 15, 1–6 | **7–9** |10–15}	qNGD algorithm				
	1.55	$2.52{\cdot}10^{-11}$	$1.77{\cdot}10^{-2}$	$5.67{\cdot}10^{-14}$	$1.81{\cdot}10^{-2}$
	1.45	$1.21{\cdot}10^{-11}$	$1.71{\cdot}10^{-2}$	$6.88{\cdot}10^{-21}$	$4.90{\cdot}10^{-6}$
	SA algorithm				
	1.55	$1.67{\cdot}10^{-2}$	0.0161	$1.31{\cdot}10^{-5}$	0.0171
{15, 15, 1–5 | **6–10** | 11–15}	qNGD algorithm				
	1.55	$2.82{\cdot}10^{-11}$	$2.92{\cdot}10^{-2}$	$2.14{\cdot}10^{-22}$	$2.70{\cdot}10^{-3}$
	1.45	$1.47{\cdot}10^{-11}$	$2.79{\cdot}10^{-2}$	$9.56{\cdot}10^{-28}$	$6.00{\cdot}10^{-4}$
	SA algorithm				
	1.55	$1.56{\cdot}10^{-2}$	0.0273	$2.94{\cdot}10^{-6}$	0.029
{15, 15, 15}	qNGD algorithm				
	1.55	$2.42{\cdot}10^{-11}$	$7.06{\cdot}10^{-2}$	$5.03{\cdot}10^{-24}$	0.119
	1.45	$1.24{\cdot}10^{-11}$	$6.65{\cdot}10^{-2}$	$7.99{\cdot}10^{-17}$	0.107
	SA algorithm				
	1.55	$1.42{\cdot}10^{-2}$	0.0658	$2.63{\cdot}10^{-8}$	0.118

As it follows from *Table 3.3*, notice that the SA algorithm code used for the spatial grid of {15, 15, 15} yields decreasing target χ^2-function value on the six orders of magnitude, whereas the 'error' parameter CP increases from 0.0658 to 0.118. A possible reason for that is the ambiguity of the inverse XRDT problem solution. At the same time, for the two spatial grids {15, 15, 1–6 | **7–9** | 10–15} and {15, 15, 1–5 | **6–10** | 11–15} ultimate values of the target χ^2-function are of the order of 10^{-22} and the 'error' parameter CP is about of $5 \cdot 10^{-5}$.

The 2D $f_{CPD}(\mathbf{r} - \mathbf{r}_0)$- cross-sections are shown for the depth levels k = 6, 8 in *Figure 3.3*.

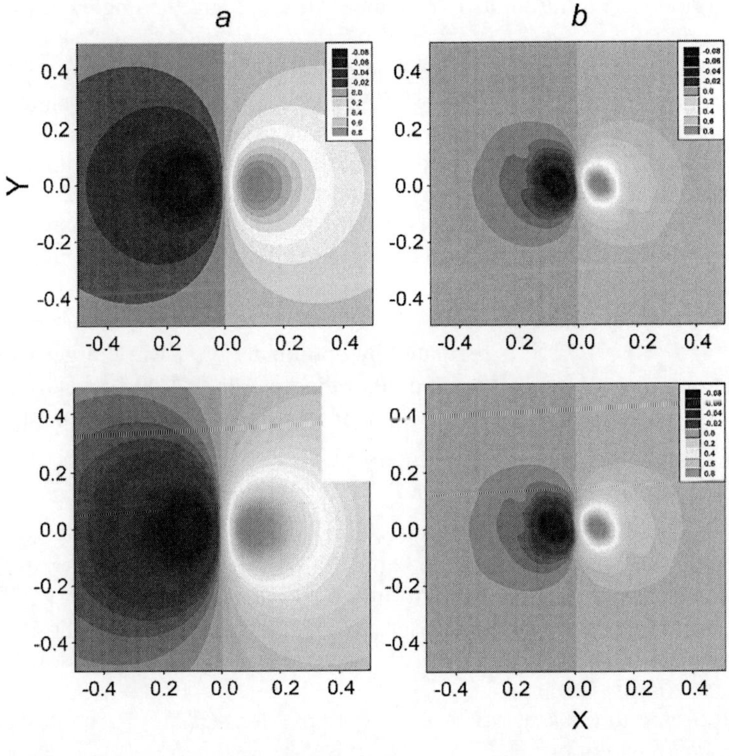

Figure. 3.3. The 2D cross-sections of the 3D $f_{CPD}(\mathbf{r} - \mathbf{r}_0)$ built at levels z: (a) z = 3/8T, (b) z= 0.5T with using the qNGD algorithm. Spatial grid net is {15, 15, 15}. The upper row displays the 2D cross sections of the 3D-defined function $f_{true}(\mathbf{r} - \mathbf{r}_0)$, $p = 1.5$.

In general, for the solution of the inverse XRDT problem, one has to launch an optimization (minimization) procedure to the XRDT target χ^2-function $\mathcal{F}\{\boldsymbol{\mathcal{P}}\}$

$$\mathcal{F}\{\boldsymbol{\mathcal{P}}\} = \frac{1}{N\{\mathrm{xy}\}}\sum_{i=1}^{N}\sum_{\{\mathrm{x}(T),\mathrm{y}(T)\}}\frac{\left(I_{h,obs}[\mathrm{x(T)},\mathrm{y(T)};\Phi_i]-I_{h,calc}[X(T),Y(T);\Phi_i,\{\boldsymbol{\mathcal{P}}\}]\right)^2}{(I_{h,obs}[X(T),Y(T);\Phi_i])^2} = min,$$

$$\mathrm{i}=1,2,\ldots\mathrm{N}. \quad (3.13)$$

Here, $I_{h,calc}[\mathrm{x}(T),\mathrm{y}(T);\Phi_i,\{\boldsymbol{\mathcal{P}}\}]$ is the diffracted intensity for the *current* vector $\boldsymbol{\mathcal{P}}$. The rectangular crystal prism is assumed as follows as $0 \leq z \leq T$, $-x/2 \leq x(T) \leq x/2, -y/2 \leq y(T) \leq y/2$. *N* is the total number of the 2D-IPs observed, $\{\Phi_i\}$ is the rotation angle set, i =1, 2,..., *N*. Vector $\boldsymbol{\mathcal{P}} = \{p, z_0, G\}$.

Further, the total number *N* assumes to be 3, and the angle set $\{\Phi_i\} = \{0°, \pm 10°\}$.

To optimize the XRDT target χ^2-function (13), one combines the qNLM algorithm (More, 1978) and Simulated Annealing (SA) algorithm (Kirkpatrick *et al.*, 1983; Aarts *et al.*, 1997) to the unique quasi-Newton-Levenberg-Marquardt-Simulated Annealing (qNLMSA) algorithm that operates in a sequential mode (*cf.* Chukhovskii, 2009). To be specific, the latter does mean that either the qNLM or SA algorithm works in a switching regime to provide a minimization procedure applied to the XRDT target χ^2-function $\mathcal{F}\{\boldsymbol{\mathcal{P}}\}$.

A transition from the *current* vector $\boldsymbol{\mathcal{P}}^{(k)}$ to the *trial* vector $\boldsymbol{\mathcal{P}}^{(k+1)}$, $\Delta\boldsymbol{\mathcal{P}}^{(k)}=\boldsymbol{\mathcal{P}}^{(k+1)} - \boldsymbol{\mathcal{P}}^{(k)}$, is regulated by optimizing the XRDT target $\mathcal{F}\{\boldsymbol{\mathcal{P}}\}$-function in a switching regime. In other words, $\boldsymbol{\mathcal{P}}^{(k+1)}$ is chosen in accordance to either $\Delta\boldsymbol{\mathcal{P}}_{qNLM}^{(k)}$- or $\Delta\boldsymbol{\mathcal{P}}_{SA}^{(k)}$-differences depending on the qNLM or the SA algorithm to be activated.

The initial value of the vector $\boldsymbol{\mathcal{P}}^{(start)} = \{p, z0, G\}$ is defined as

$$=\mathcal{P}_j^{(true)}\{1 + \text{A}\mathrm{Random}[-1,1]\}, \quad (3.14)$$

where A is the amplitude of the $\mathrm{Random}[-1,1]$, which provides pseudorandom real numbers uniformly distributed in the range of [– 1, 1]. In the further calculations, the set $\{\text{A}\} = \{0.1, 0.2, 0.3, 0.4\}$ is chosen for approving the qNLMSA algorithm code in its application to solve the inverse XRDT problem.

Referred to the *k*-iteration, the Figure of Merit (FOM) R_k controls the calculation evolution

$$R_k = \frac{\sum_{\Phi}\sum_{\{\mathrm{x(T)},\mathrm{y(T)}\}}\left|A_{h,obs}[\mathrm{x(T)},\mathrm{y(T)};\Phi]-A_{h,calc}[\mathrm{x(T)},\mathrm{y(T)};\Phi,\{\boldsymbol{\mathcal{P}}^{(k)}\}]\right|}{\sum_{\Phi}\sum_{\{\mathrm{x(T)},\mathrm{y(T)}\}}A_{h,obs}[\mathrm{x(T)},\mathrm{y(T)};\Phi]}, \quad (3.15)$$

where the following notations are introduced

$$A_{h,obs}[\mathrm{x(T)},\mathrm{y(T)};\Phi] = I_{h,obs}^{1/2}[\mathrm{x(T)},\mathrm{y(T)};\Phi],$$

$$A_{h,calc}[\mathrm{x(T)},\mathrm{y(T)};\Phi] = I_{h,calc}^{1/2}[\mathrm{x(T)},\mathrm{y(T)};\Phi,\{\boldsymbol{\mathcal{P}}^{(k)}\}]. \quad (3.15a)$$

To be definite, the flow equation that guides the XRDT target χ^2-function $\mathcal{F}\{\boldsymbol{\mathcal{P}}\}$ processing takes the form

---

– Assign the initial vector $\boldsymbol{\mathcal{P}}^{(\mathrm{start})} = \{p, \mathrm{z}_0, G\}$;
– Assign the switch number m, $0 \leq m \leq M$, M=10;
Do [
Do [Evaluate differential $\Delta\boldsymbol{\mathcal{P}}_{qNLM}^{(\boldsymbol{m},\boldsymbol{k})}$ for the activated qNLM algorithm;
Evaluate FOM $R_{m,k}$;
Terminate the $\mathcal{F}\{\boldsymbol{\mathcal{P}}^{(m,k)}\}$ processing
if $\frac{|\mathcal{F}\{\boldsymbol{\mathcal{P}}^{(m,k+1)}\} - \mathcal{F}\{\boldsymbol{\mathcal{P}}^{(m,k)}\}|}{\mathcal{F}\{\boldsymbol{\mathcal{P}}^{(m,k)}\}} < 10^{-10}$, $\{k, 1, K\}$];
Terminate the $\mathcal{F}\{\boldsymbol{\mathcal{P}}^{(m,K^*)}\}$ processing
if $\mathcal{F}\{\boldsymbol{\mathcal{P}}^{(m,K^*)} \leq 10^{-10}\} \cup R_{m,K*} \leq 10^{-6}$ for k=K^*, $K^* \leq K$,
$\{m, 0, M\}$];

It should be noted that the idea of hybridization of some computer algorithms and thereby improving the convergence of solving inverse problems was previously discussed and used in the works (Chukhovskii, 2009; Martin & Otto, 1996; Huynh & Roy, 2009; Leonetti, Kormushev, Sagratella, 2012; Wu *et al*., 2014). Authors (Kernighan & Lin, 1970) reported that combining the SA and Newton-types algorithms has improved over the local search of about 8.9% for the partitioning of sparse random graphs of average degree equal to 5. Such the idea of the qNLMSA algorithm run-through in the case of the *inverse* XRDT problem seems to be effective and opens 'a door' for escaping of the $\mathcal{F}\{\boldsymbol{\mathcal{P}}\}$-function from the local minimum traps if they are. To be specific, should the observed 2D-IPs have a good quality image contrast, the XRDT $\mathcal{F}\{\boldsymbol{\mathcal{P}}\}$ – function has an excellent chance to tend to the global minimum by following its evolutionary path.

Loosely speaking, the use of the qNLMSA algorithm aims to overcome the inverse XRDT problem processing difficulties.

As is shown underneath, the qNLMSA algorithm works well, avoiding abrupt situations of slipping the χ^2-function $\mathcal{F}\{\boldsymbol{\mathcal{P}}\}$ to local minima leads to the processing stagnation, especially if the parameters of the initial vector $\boldsymbol{\mathcal{P}}^{(start)}$ are relatively far away from the *true* one $\boldsymbol{\mathcal{P}}^{(true)}$.

3.4. Solution of the X-ray diffraction tomography problem using the quasi-newton-levenberg-marquardt - simulated annealing algorithm

To avoid some cumbersome mathematical formulae, we confine ourselves to the case of the σ-polarized X-ray wavefield propagating through a crystal Si(111) oriented in the exact Bragg position, when the angular deviation parameter w=0.

Processing of the XRDT $\mathcal{F}\{\mathcal{P}\}$-function is carried out in the case of the noise-free and the Gaussian 3%-noise 2D-IPs, respectively. The corresponding 2D-IPs $I_{h,obs}[\mathrm{x(T), y(T)}; \Phi]$ are calculated for three values of the rotation angle Φ= {0°, ±10°}, the actual vector $\mathcal{P}^{(\mathrm{true})} = \{1.5, 0.5\mathrm{T},\ 1.8,\}$, T = π, -π ≤ x(T) ≤ π, -π ≤ y(T) ≤ π.

The latter are depicted in Figs. 3.4 for the noise-free (left column) and the Gaussian noise 2D-IPs (right column), respectively. The Gaussian noise 2D-IPs are obtained by adding the Gaussian noise to the noise-free 2D-IPs (Marsaglia& Tsang, 1984).

As it follows from *Figure 3.4*, the image contrast maximum is of the order of 13.6 % in *Figure 3.4a* (Φ = -10^0) and *Figure 3.4c* (Φ = 10^0), whereas it is of the charge of 15.9 % in *Figure 3.4b* (Φ = 0^0). The artificial 2D-IPs represent the double-form images of the CPD due to the dipole-kind behavior of the 2D f_{CPD} (**r**-**r**$_0$) cross-sections.

Next, processing of the XRDT $\mathcal{F}\{\mathcal{P}\}$ -function has been launched using the qNLMSA algorithm code in the switching regime described above. Eventually, the corresponding evaluations were carried out for amplitudes $\mathbb{A} = \{0.1, 0.2, 0.3, 0.4\}$ of the initial vectors $\mathcal{P}^{(start)}$. It is essential that in the case of decoding data of the noise-free 2D-IPs, the retrieval vectors $\mathcal{P}^{(end)}$ coincide with the actual vector $\mathcal{P}^{(true)}$ for all the chosen amplitudes $\{\mathbb{A}\}$ and in turn, it does mean that the absolute (global) minimization of the XRDT $\mathcal{F}\{\mathcal{P}\}$ - function has been achieved. In the case of the noise-free 2D-IPs, how the qNLMSA algorithm works have been illustrated in *Table 3.4*. The corresponding sequential stages of the qNLMSA processing of the XRDT $\mathcal{F}\{\mathcal{P}\}$-functions are displayed in *Figure*. 3.5.

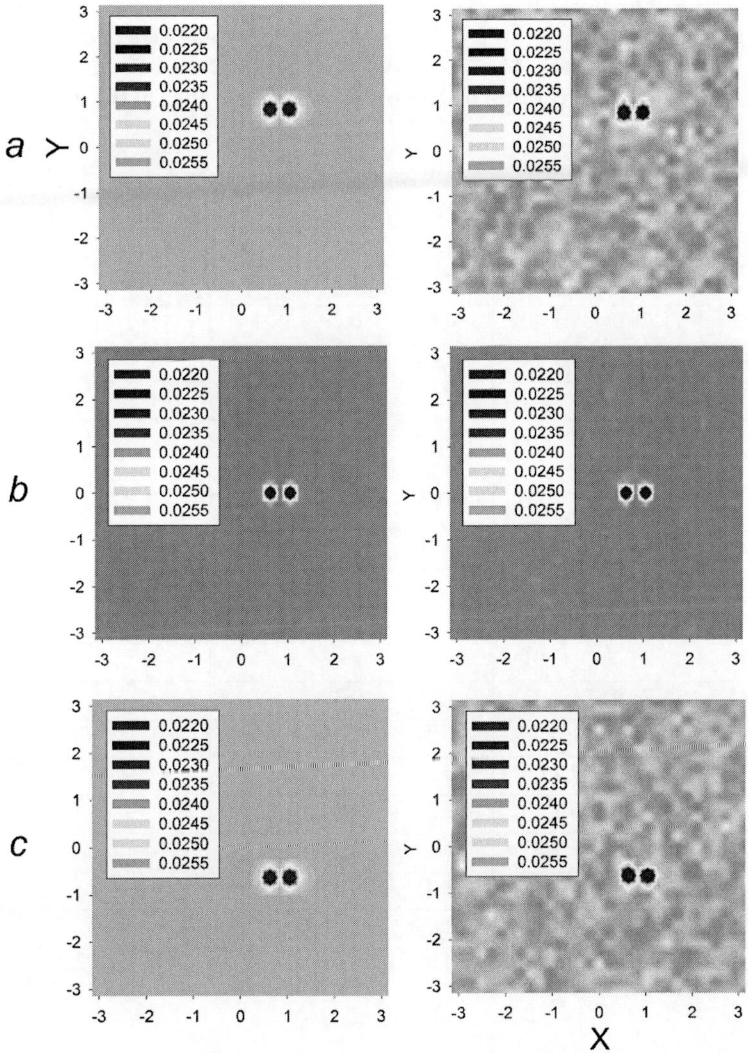

Figure 3.4. The simulated noise-free (left column) and the Gaussian 3%-noise (right column) 2D-IPs. The rotation angle Φ is equal to: (a) -10°, (b) 0°, (c) 10°.

Table 3.4. **The noise-free 2D-IPs. The spatial grid {31, 31, 21} referred to the rectangular sample prism is used. The voxel sizes coordinates along the (x, y, z) are Δx= π/30, Δy= π/30, Δz= π/20. Event number No is given in the first column of *Table 3.4.* *m* is the number of the qNLMSA code switches.**

No	$\mathcal{P}^{(start)}$	$\mathcal{F}\{\mathcal{P}^{(start)}\}$ *10^5	$R^{(start)}$	$\mathcal{P}^{(end)}/\mathcal{P}^{(true)}$	$\mathcal{F}\{\mathcal{P}^{(end)}\}$ *10^-10	$R^{(end)}$ *10^-6	m
1	(1.12;0.554;2.42)	5.48	6.85	1	1	1	0
2	(1.26;0.567;2.05)	2.95	2.21	1	1	1	0
3	(1.71;0.485;1.40)	0.62	0.38	1	1	1	0
4	(1.79;0.555;1.16)	2.04	0.44	1	1	1	0
5	(2.19;0.310;1.65)	1.93	0.49	1	1	1	0
6	(1.18;0.595;2.01)	3.27	3.74	1	1	1	0
7	(1.38;0.431;1.70)	0.24	0.51	1	1	1	0
8	(2.03;0.453;2.06)	0.07	0.47	1	1	1	2
9	(1.46;0.433;1.72)	0.25	0.13	1	1	1	4
10	(1.89;0.463;1.12)	1.03	0.46	1	1	1	2
11	(2.07;0.371;2.46)	2.32	0.48	1	1	1	2
12	(2.26;0.451;2.01)	0.09	0.42	1	1	1	4
13	(1.73;0.671;2.23)	1.75	0.37	1	1	1	4
14	(1.49;0.330;.1.33)	1.87	0.19	1	1	1	2
15	(1.16;0.387;2.06)	0.58	4.52	1	1	1	4
16	(1.90;0.630;2.30)	2.30	0.44	1	1	1	4
17	(2.06;0.582;1.51)	2.01	0.47	1	1	1	2
18	(1.78;0.561;1.33)	2.03	0.42	1	1	1	8
19	(2.29;0.626;1.95)	2.08	0.48	1	1	1	4
20	(2.08;0.397;1.99)	0.05	0.43	1	1	1	2

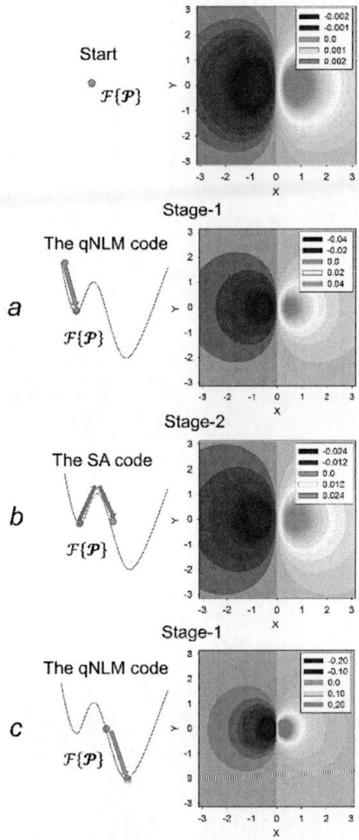

Figure 3.5. Schematic of the qNLMSA algorithm code:
Start $\mathcal{P}^{(start)}$={1.70,0.43,1.38}, $\mathcal{F}\{\mathcal{P}^{(start)}\}$=0.24*10^{-5}, $\text{R}_k{}^{(start)} = 0.51$,
(a) $\mathcal{P}^{(a)}$={1.49,0.45,1.82},$\mathcal{F}\{\mathcal{P}^{(a)}\}$=0.68*10^{7}, $\text{R}_k{}^{(a)} = 0.12$,
(b) $\mathcal{P}^{(b)}$={1.60, 0.41, 2.17}, $\mathcal{F}\{\mathcal{P}^{(b)}\}$ = 0.79*10^{-7}, $\text{R}_k{}^{(b)} = 0.40$,
(c) $\mathcal{P}^{(c)}$={1.50, 0.50, 1.80},. $\mathcal{F}\{\mathcal{P}^{(c)}\}$ = 0.34*10^{-25}, $\text{R}_k{}^{(c)}$ = 0.27*10^{-9}.
Vector $\mathcal{P}^{(end)} = \mathcal{P}^{(c)}$. The curved line depicts the behavior of the XRDT χ^2-function $\mathcal{F}\{\mathcal{P}\}$. The 2D $f_{\text{CPD}}(\mathbf{r} - \mathbf{r}_0)$ –function cross-sections are shown in the right column at the level z_0/T=0.

The results of processing the corresponding Gaussian 3%-noise 2D-IPs are listed in *Table 3.5*. From *Table 3.5,* it follows that on average, the retrieval vectors $\mathcal{P}^{(end)}$ are around the vector $\mathcal{P}^{(true)}$ with an accuracy of up to 5% while the FOM $\text{R}^{(end)}$-values are lying in limits of 10%÷15%.

***Table 3.5*. The Gaussian 3%-noise 2D-IPs. The spatial grid {31, 31, 21} referred to the rectangular sample prism is used. The voxel sizes coordinates along the (x, y, z) are $\Delta x = \pi/30$, $\Delta y = \pi/30$, $\Delta z = \pi/20$. The event number No is given in the first column of *Table 3.5*, *m* is the number of the qNLMSA code switches.**

No	$\mathcal{P}^{(start)}$	$\mathcal{F}\{\mathcal{P}^{(start)}\}*10^5$	$R^{(start)}$	$\mathcal{P}^{(end)}$	$\mathcal{F}\{\mathcal{P}^{(end)}\}*10^5$	$R^{(end)}$	m
10	(1.89;0.463;1.12)	1.03	0.46	(1.47;0.488;1.78)	3.51 $*10^{-6}$	0.13	10
2	(1.26;0.567;2.05)	2.95	2.21	(1.46;0.498;1.80)	2.41 $*10^{-6}$	0.08	10
3	(1.71;0.485;1.40)	0.62	0.38	(1.51;0.497;1.82)	3.07 $*10^{-6}$	0.11	10
19	(2.29;0.626;1.95)	2.08	0.48	(1.51;0.499;1.77)	2.27 $*10^{-6}$	0.07	10
5	(2.19;0.310;1.65)	1.93	0.49	(1.52;0.485;1.78)	3.29 $*10^{-6}$	0.10	10

As the routine, straightforward calculations for recovering the 3D-defined function $f_{\mathrm{CPD}}(\mathbf{r}-\mathbf{r}_0)$, indicate that collecting a large number, *e.g.* $\Phi_i = 2^0(i-11)$, $(i = 1, 2, \dots, 21)$, for the tilted Gaussian 3%-noise 2D-IP data set does not change or affect a conclusion concerning good quality of the 3D-defined function $f_{\mathrm{CPD}}(\mathbf{r}-\mathbf{r}_0)$. Moreover, enlarging the number of 2D-IPs may enhance the accuracy of the vector $\boldsymbol{\mathcal{P}}^{(end)}$ retrieval. The latter can be essential to improve the quality of the filtered XRDT 2D-IPs data involved in solving the inverse XRDT problem.

3.5. Conclusion

On the example of the Coulomb-types point defect in a crystal, one managed to show that in the theoretical framework, in particular, a semi-kinematical solution of the basic Takagi-Taupin equations, the qNLMSA algorithm code provides successful processing of the XRDT χ^2-function to solve the inverse XRDT problem in question.

As to the theoretical framework, processing the XRDT χ^2-function based on the semi-kinematical approach to the XRDT problem solution, which relates to the direct defect image contrast on the corresponding 2D-IPs,

The advantage of the qNLMSA algorithm is its capacity to provide the absolute (global) minimum of the XRDT χ^2-function using the noise-free (filtered) 2D-IPs data in a proper way.

Notice that in the case of the minor size defects, a high-level noise of the CCD detectors may significantly complicate a situation since the noise-filtering methods that existed for a moment may not be effective. Nevertheless, commercial high-resolution CMOS detectors have appeared that, unlike the CCD devices, do not require cooling and have a lower intrinsic noise level (Hamamatsu Photonics Company Communication, 2019).

Concerning a general solution of the inverse XRDTT problem, how both the mathematical framework and the qNLMSA algorithm presented work in the cases of other kinds of crystal lattice defects remains still open for future research.

Based on the results of processing the XRDT χ^2-function for the CPD, one reasonably supposes that to obtain direct information about elastic displacement fields around small-size crystal lattice defects, particularly interstitial and vacancy-types clusters, small dislocation loops, *etc.* employing the *ab initio* XRDT technique has good prerequisites.

References

Aarts, E., Korst, J. & van Laarhoven, P. *"Simulated annealing," in Local Search in Combinatorial Optimization*. Aarts, E. H. L. and Lenstra, J. K. (Eds.), John Wiley and Sons, Ltd., 1997, pp. 91–120.

Asadchikov, V., Besedin, I, Buzmakov, A. Chukhovskii, F., Zolotov, D. (2014) *Acta Crystallogr. A* **70**, C1132.

Asadchikov, V., Buzmakov, A., Chukhovskii, F., Dyachkova, I., Zolotov, D., Danilewsky, A., Baumbach, T., Bode, S., Haaga, S., Daniel Hänschke, D., Kabukcuoglu, M., Balzer, M., Caselle, M., & Suvorov, E. (2018). *J. Appl. Cryst.*, **51**, 1616–1622.

Authier A., Lang A.R. (1964). *J. Appl. Phys.*, **35**, 1956-1962.

Authier, A. (2001). *Dynamical Theory of X-ray Diffraction*. New York: Oxford Univ. Press.

Benediktovitch, A., Feranchuk, I. & Ulyanenko, A. (2013). *Theoretical Concepts of X-ray Nanoscale Analysis*. Springer, Heidelberg.

Besedin, I. S., Chukhovskii, F. N. &Asadchikov, V. E. (2014). *Crystallogr. Rep.*, **59**, 323–330.

Bowen D. K., Tanner B. K. *High-Resolution X-Ray Diffractometry and Topography*. (1998). New York: Taylor & Francis.

Chukhovskii, F.N. (2009). *Acta Cryst.*, **A65**, 39–45.

Chukhovskii, F. N., Konarev, P. V. & Volkov, V. V. (2019) *Scientific Reports* **9**, 14216.

Chukhovskii, F. N., Konarev, P. V. & Volkov, V. V. (2020) *Acta Crystallogr. A* **76**, 16-25.

Danilewsky, A. N., Wittge, J., Croell, A., Allen, D., McNally, P., Vagovič, P., dos Santos Rolo, T., Li, Z., Baumbach, T., Gorostegui-Colinas, E., Garagorri, J., Elizalde, M. R., Fossati, M. C., Bowen, D. K. & Tanner, B. K. (2011). *J. Cryst. Growth*, **318**, 1157–1163.

Dennis, J., Gay, D., Welsch, R. (1981) *ACM Trans. Math. Soft*. **7**, 348-368.

Epelboin. Y. (1975) *Acta Crystallogr. A* **31**, 591-600.

Gill, P.E., Murray, W. and Wright, M.H. (1981) *Practical Optimization*, Academic Press, London.

Guinier, A. (1956). *Theorie et technique de la radiocristallographie*. Paris: Dunod.

Hänschke, D., Helfen, L., Altapova, V., Danilewsky, A. & Baumbach, T. (2012). *Appl. Phys. Lett.*, **101**, 244103.

Hänschke, D., Danilewsky, A. N., Helfen, L., Hamann, E. & Baumbach, T. (2017). *Phys. Rev. Lett.*, **119**, 215504.

Huynh, V.A., Roy, N. (2009) *Robotics and Automation, IEEE International Conference 2009*, 2851-2858.

Kawado, S., Taishi, T., Iida, S. *et al.* (2004) *J. Synchrotron Radiat.* **11**, 304-308.

Kernighan, B.W., Lin, S. (1970). *Bell System Tech. J.*, **49**, 291-307.

Kirkpatrick S., Gelatt C. D., Vecci M. P. (1983). *Science*, **220**, 671-680.

Konarev, P. V., Chukhovskii, F.N., Volkov, V.V. (2019) *Crystallogr. Rep.*, **64**(2), 191-200.

Leonetti, M., Kormushev, P., Sagratella, S. (2012) *Cybernetics and information technologies*, **12**, 53-65.

Ludwig, W., Clötens, P., Härtwig, J., Baruchel, J., Hamelin, B. & Bastie, P. (2001). *J. Appl. Cryst.*, **34**, 602–607.

Marsaglia, G., Tsang, W.W. (1984). *SIAM Journal on Scientific and Statistical Computing*, **5**, 349-359.

Martin, O.C., Otto S.W. (1996) *Annals of Operations Research*, **63**, 57-75.

Metropolis, N., Rosenbluth, A.W., Rosenbluth, M.N. *et al.* (1953) *Chem. Phys.* **21**, 1087-1092.

More J.J. (1978). *The Levenberg-Marquardt algorithm, implementation, and theory*. Berlin and New York: Springer-Verlag, Springer Lecture Notes in Mathematics, No. 630, ed. G.A. Watson.

Pavlov, K. M. and Punegov, V. I. (2000). *Acta Cryst.* A**56**, 227-234.

Seregin, A. Yu., Prosekov, P. A., Chukhovskii, F.N., Volkovsky, Yu. A., Blagov, A.E., and Kovalchuk, M. V. (2019). *Crystallogr. Rep.*, **64(4)**, 545-552.

Takagi, S. (1962). *Acta Cryst.*, **15**, 1311–1312.

Takagi, S. (1969). *J. Phys. Soc. Jpn*, **26**, 1239–1253.

Tanner B. K., Phil M. A. (1966). *X-Ray Diffraction Topography*. New York: Pergamon Press.

Taupin, D. (1964). *Bull. Soc. Fr. Miner. Crist.*, **87**, 469-511.

Wu, W., Liang, J., Yao, X, Liu, B. (2014) *Mathematical Problems in Engineering*, **2014**, 151394.

Zolotov, D. A., Buzmakov, A. V., Elfimov, D. A., Asadchikov, V. E. & Chukhovskii, F. N. (2017). *Crystallogr. Rep.*, **62**, 20–24.

CHAPTER FOUR

THE X-RAY HIGH-RESOLUTION REFLECTOMETRY

4.1. Introduction

Determining the multi-layer structure (MLS) parameters such as the thickness, complex refraction index of each layer, the surface roughness, and the layer interface roughness using the high-resolution X-ray reflectometry (HRXR) technique is of particular interest in applied physics and technology. In general, while methods such as surface X-ray diffraction, absorption X-ray spectroscopy, and X-ray fluorescence spectroscopy all work well, each of them suffers physical limitations since none are sensitive to all the MLS parameters together. In general, while methods such as surface X-ray diffraction, absorption X-ray spectroscopy, and X-ray fluorescence spectroscopy all work well, each of them suffers physical limitations since none are sensitive to altogether of the MLS parameters. In the present, the state-of-art progress in the MLS studies with using the HRXR technique has become evident due to the pioneering works (Cowley & Ryan, 1987; Sinha *et al.*, 1988; Holy *et al.*, 1999; Baake et al., 2006; Bridou *et al.*, 2006; Lemoine *et al.*, 2006; Hodroj *et al.*, 2006). The theoretical model of the grazing-incidence X-ray specular scattering by an MLS was first developed by Parratt (1954) based on the partial Fresnel reflections by each layer of the MLS. Afterward, all the partial Fresnel transmission and reflection coefficients have been modified by the relevant Debye-Waller factors first introduced by Nevot & Croce (1980). Both the transmission and reflection Debye-Waller-types factors are due to the MLS surface and interface roughness, and they effectively influence the grazing-incidence X-ray specular scattering output. It has become clear that the most advanced method for detecting the MLS parameters is to decode the 1D HRXR θ-scan data (see *Figure 4.1*; θ is the grazing-incidence angle of the X-ray plane-wave). The MLS roughness model (Nevot & Croce, 1980) seemed adequate for retrieving the MLS parameters using the HRXR technique.

As is well known, various iterative computer algorithms exist for solving inverse physical problems. Amid them are frequently named the Newton and Gradient Descend ones (see, *e.g.*, Press *et al.*, 1992; Eadie *et al.*, 1971; Gill *et al.*, 1981; Golberg, 1989). As to the quasi-Newton-Levenberg-Marquardt (qNLM) algorithm, the latter is a hybridization of the Newton and Gradient Descend algorithms.

There is virtue in the computer iterative methods for solving the inverse physical problem that all they are based upon the error functional function $F\{\mathbf{P}\}$ optimization in a least-square fashion over the multidimensional vector **P** *in search*.

Hereafter, the qNLM algorithm concept is pushed one step further as an advanced prerequisite to reaching the absolute (global) minimum of the HRXR function $F\{\mathbf{P}\}$ even the initial vector **P** is chosen rather far away from the true one. Remind that by definition, the absolute (global) minimum of the functional error $F\{\mathbf{P}\}$ refers to its zero-value with the true MLS vector **P**.

Such as the Newton-Gradient Descend (NGD) algorithm has been proposed in (Chukhovskii, 2009). Being convergent in a statistical sense, it works in the alternative mode and provides a plausible solution to the inverse HRXR problem based on the modified Parratt relationships. By using the artificial HRXR θ -scan data, the NGD algorithm has been probated for the three-layer structure (Chukhovskii, 2009). For reference, using a short wavelength of the incident X-rays, within the range 0.1-1 nm, allows detecting all the three-layer structure parameters, including the nanoscale surface and/or interface roughness.

4.2. The Modified Parratt Recurrent Relationships

Before proceeding further, we clarify the mathematical formalism to describe the grazing-incidence X-ray specular scattering from the MLS in terms of the Parratt recurrent relationships (*cf.* Nevot & Croce, 1980; Bushuev & Sutyrin, 2000).

In the case of the ideally flat surface and interfaces of the MLS, the Parratt recurrent relationships take the form (*cf.* schematic in *Figure 4.2*)

$$R_{n-1}^{(id)} = \frac{r_{n-1,n}^{(0)} + \left(-r_{n-1,n}^{(0)} \cdot r_{n,n-1}^{(0)} + t_{n-1,n}^{(0)} \cdot t_{n,n-1}^{(0)}\right) \cdot R_n^{(id)} \cdot g_n}{1 + r_{n-1,n}^{(0)} \cdot R_n^{(id)} \cdot g_n}, . \quad (4.1)$$

$(n = 1, 2, ..., N)$

Here $r^{(0)}_{n-1,n}$ is the Fresnel reflection coefficient of the plane-wave $\exp[i\boldsymbol{k}^{(n-1)}\boldsymbol{r}]$ that is propagating through the (n-1)-th layer ($r^{(0)}_{n,n-1} = -r^{(0)}_{n-1,n}$)

$$r^{(0)}_{n-1,n} = \frac{k_z^{(n-1)} - k_z^{(n)}}{k_z^{(n-1)} + k_z^{(n)}}. \tag{4.2}$$

The exponential factor g_n in equation (4.1) describes the phase shift due to the plane-wave propagation through the n-th layer

$$g_n = \exp\left(2ik_z^{(n)}T^{(n)}\right) \tag{4.3}$$

The standard component of the wave vector $k_z^{(n)}$ within the nth layer is defined as

$$k_z^{(n)} = (2\pi/\lambda)\cdot(\sin^2\theta + \chi^{(n)})^{1/2}, \tag{4.4}$$

where λ is the wavelength of the X-ray plane-wave, $T^{(n)}$ is the nth layer thickness, and $\chi^{(n)}$ is the zero Fourier component of the electric susceptibility $\chi^{(n)}(\mathbf{r})$ of the nth layer ($\chi^{(0)} = 0$ for a vacuum).

Accordingly, the Fresnel transmission coefficient $t^{(0)}_{n-1,n}$ is equal to

$$t^{(0)}_{n-1,n} = \frac{2k_z^{(n-1)}}{k_z^{(n-1)} + k_z^{(n)}}. \tag{4.5}$$

Note that the following identity $-r^{(0)}_{n-1,n}\cdot r^{(0)}_{n,n-1} + t^{(0)}_{n-1,n}\cdot t^{(0)}_{n,n-1} \equiv 1$ holds.

In principle, using the fundamental recurrent relationships (4.1) and equations (4.2)-(4.5) and bearing in mind that $R_N^{(id)} = 0$ and $R_{N-1}^{(id)} = r^{(0)}_{N-1,N}$, the self-consistent procedure for calculating the HRXR $|R_0^{(id)}(\theta)|^2$-scan for the ideally planar MLS may be developed.

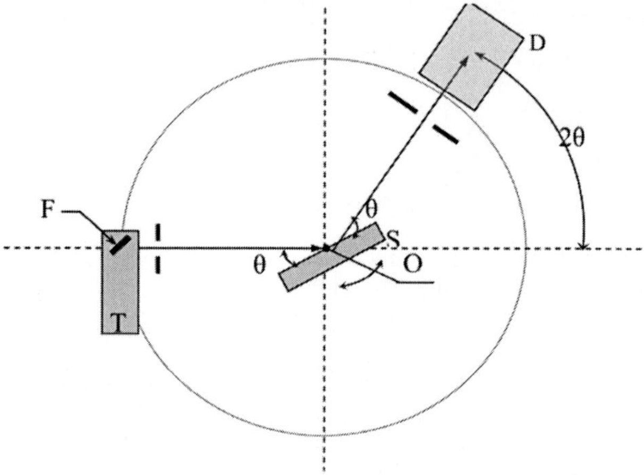

Figure 4.1. The HRXR geometry. TF is the X-ray source, S is the MLS sample, D is the detector position. θ is the X-ray grazing-incidence angle (the scale is exaggerated).

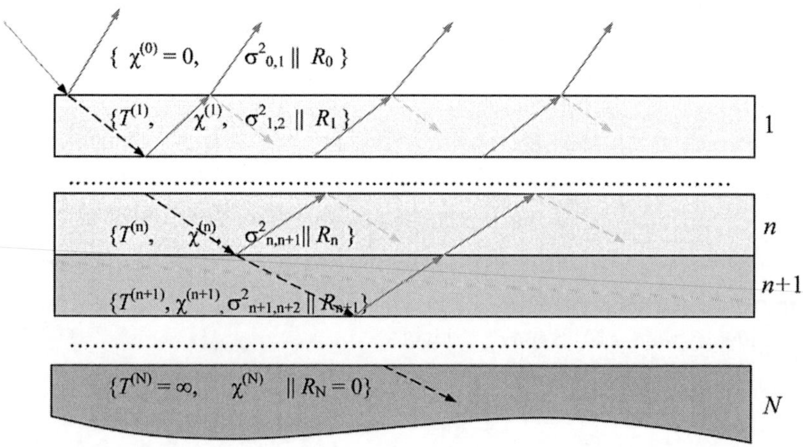

Figure 4.2. Schematic of an MLS. $T^{(n)}$ and $\chi^{(n)}$ are the thickness and zero-order electric susceptibility of the *n*-th layer, and the $\sigma^2_{n-1,n}$ is the mean-square roughness of the (n - 1, n) interface.

Accepting the concept of the continuous wave fields and their derivatives along the normal to each random interface $z_{n-1,n} = u_{n-1,n}(x, y)$ that is the layer boundary (each function $u_{n-1,n}(x, y)$ is smooth and single-valued). In the case of the real MLS one needs to modify the ideal Fresnel coefficients $r_{n-1,n}^{(0)}$ and $t_{n-1,n}^{(0)}$ by the corresponding Debye-Waller factors (Nevot & Croce, 1980)

Accordingly, the following solutions for the Fresnel coefficients $r_{n\,-1,n}$ and $t_{n\,-1,n}$

$$r_{n-1,n} = \exp\left(2ik_z^{(n-1)} u_{n-1,n} \right) r_{n-1,n}^{0} , \qquad , \qquad (4.6)$$
$$t_{n-1,n} = \exp\left[2i \left(k_z^{(n-1)} - k_z^{(n)} \right) u_{n-1,n} \right] \cdot t_{n-1,n}^{(0)}$$

where $\sigma_{n-1,n}^2$ is the mean-square roughness of the $(n\,-\,1,\,n)$-th layer interface.

Notice that the above derivation of the static exponential factors [equation (4.6)] yields the same expression for the Debye-Waller transmission term. The latter is similar to one obtained in (Nevot & Croce, 1980), whereas the Debye-Waller reflection factor differs from the corresponding Nevot & Croce's factor.

Accordingly, in the case of the $(n\,-\,1,\,n)$-th layer interface, the ideal Fresnel coefficients are modified as follows

$$\hat{\boldsymbol{r}}_{n-1,n} = f_{n-1,n}^{(r)_{n-1,n}} \boldsymbol{r}_{n-1,n}^{(0)} , \qquad (4.7)$$
$$\hat{\boldsymbol{t}}_{n-1,n} = f_{n-1,n}^{(t)_{n-1,n}} \boldsymbol{r}_{n-1,n}^{(0)} .$$

Then, the modified Parratt recurrent relationships take the form (*cf.* (Bushuev & Sutyrin,2000)

$$R_{n-1} = \frac{\hat{\boldsymbol{r}}_{n-1,n} + \left(-\hat{\boldsymbol{r}}_{n\ 1,n} \hat{\boldsymbol{r}}_{n,n\ 1} + \hat{\boldsymbol{t}}_{n-1,n} \hat{\boldsymbol{t}}_{n,n-1} \right) R_n g_n}{1 + \hat{\boldsymbol{r}}_{n-1,n} R_n g_n} , . \qquad (4.8)$$

$$(n = 1, 2, \ldots, N)$$

Notice that the identity is not satisfied due to the Debye–Waller factors.

The relationship (4.8) is crucial for evaluating the intensity $|R_0^{(ex)}(\theta)|^2$ that describes the grazing-incidence X-ray specular scattering by the imperfect MLS.

Based on equations (4.1) - (4.8), in the case of the ideal and imperfect three-layer structures, the corresponding HRXR angular $-\,ln[\ |R_0^{(id)}(\theta)|^2]$ - and $-\,ln[\ |R_0^{(ex)}(\theta)|^2]$ -scans are simulated in a logarithmic scale (see *Figure 4.3a* and *Figure 4.3b*).

In the next Section, devoted to the inverse HRXR problem solution, the HRXR angular scan depicted in *Figure 4.3b* is used for determining all the three-layer imperfect structure parameters, including the surface and interface roughness.

4.3. Solution of the X-ray high-resolution reflectometry problem

Let us introduce the parameter vector $\{\boldsymbol{P}\}=\sum_{n=1}^{N}\{P_n\}$, $P_n=(T^{(n-1)},\ \chi^{(n-1)},\ \sigma^2_{n-1,n})$, which characterizes the imperfect MLS in question, and one needs to generate the processing procedure to minimize the HRXR error functional function

$$F\{\boldsymbol{P}\}=J^{-1}\sum_{j=1}^{J}\left[\ \left|R_0^{(ex)}(\theta_j)\right|^2-\left|R_0(\theta_j;\{P\})\right|^2\ \right]\Big/\left|R_0^{(ex)}(\theta_j)\right|^2 \tag{4.11}$$

To process the error-functional function $F\{\boldsymbol{P}\}$, one converts the Newton and GD algorithms to the combined NGD algorithm that operates in an alternative regime (Chukhovskii, 2009). In principle, each of the Newton and Gradient Descend ones is an adequate tool for attempting to solve the inverse HRXR problem. As to the combined NGD algorithm, it operates in an alternative fashion. The latter does mean that within each k-iteration process, the transition from the $F\{\boldsymbol{P}^{(k)}\}$-value to the $F\{\boldsymbol{P}^{(k+1)}\}$-value is governed by the switch criterion in working either the Newton

Namely, the choice of the difference $\Delta\boldsymbol{P}^{(k)}=\boldsymbol{P}^{(k+1)}-\boldsymbol{P}^{(k)})$ is made as follows:

$\Delta\boldsymbol{P}^{(k)}=\Delta\boldsymbol{P}^{(k)}_{\mathrm{NA}}\cap\Delta\boldsymbol{P}^{(k)}_{\mathrm{GA}}$ (the notation a ∩ b denotes either a or b); (4.12a)

$\Delta\boldsymbol{P}^{(k)}_{\mathrm{NA}}\propto-\hat{\mathbf{H}}[\boldsymbol{P}^{(k)}]\mathbf{A}[\boldsymbol{P}^{(k)}]$ (Newton's algorithm);

$\Delta\boldsymbol{P}^{(k)}_{\mathrm{GA}}\propto-\breve{\mathbf{I}}\mathbf{A}[\boldsymbol{P}^{(k)}]$ (the GD algorithm). (4.12b)

Here, $\mathbf{A}[\boldsymbol{P}]$ is the gradient vector, the first-order partial derivative $A_i[\boldsymbol{P}]=\partial F\{\boldsymbol{P}\}/\partial P_i$. $\hat{\mathbf{H}}[\boldsymbol{P}]$ is the Hessian matrix. The latter is the inverse matrix to the symmetric matrix of the second-order partial derivatives of the error function $F\{\boldsymbol{P}\}$ $\hat{\mathbf{H}}[\boldsymbol{P}]=\hat{\mathrm{G}}^{-1}[\boldsymbol{P}]$, $\mathbf{G}_{i,j}[\boldsymbol{P}]=\partial^2F[\boldsymbol{P}]\big/\partial P_i\cdot\partial P_j$ and $\breve{\mathbf{I}}$ is the unit matrix.

The flow equation of the NGD algorithm looks as follows:

Do [

Do [

– Assign the initial vector $\boldsymbol{P}$ as

$P_i^{(start)}=P_i^{(true)}(1+\Lambda\cdot Random[-1,\ 1])$,

where Λ is the deviation amplitude, Λ < 1, the function Random[-1, 1] yields uniformly distributed pseudorandom real numbers in the range [-1, 1];

– Evaluate the $F\{\boldsymbol{P}^{(k)}\}$, the gradient vector $\mathbf{A}[\boldsymbol{P}^{(k)}]$ and the second-order-partial derivatives matrices $\hat{\mathbf{G}}[\boldsymbol{P}^{(k)}]$, $\hat{\mathbf{H}}[\boldsymbol{P}] = \hat{\mathbf{G}}^{-1}[\boldsymbol{P}^{(k)}]$;

– Evaluate the difference $\Delta\boldsymbol{P}^{(k)}$ for both the Newton and GD algorithms, and choose the lower value of the $F\{\boldsymbol{P}^{(k+1)}\}$;
– Fix the $F\{\boldsymbol{P}^{(k+1)}\}$ at the vector $\boldsymbol{P}^{(k+1)}$;
– Terminate the iteration process if the $F\{\boldsymbol{P}^{(k+1)}\}$-value $\leq 10^{-9}$ (the number of 10^{-9} is accepted as if it provides the global minimum of $F\{\boldsymbol{P}^{(\text{true})}\}$);
– Output the final set of $\{\boldsymbol{P}^{(\text{true})}\}$, $\{k,1,K\}]$, $\{n, 1, N\}]$.

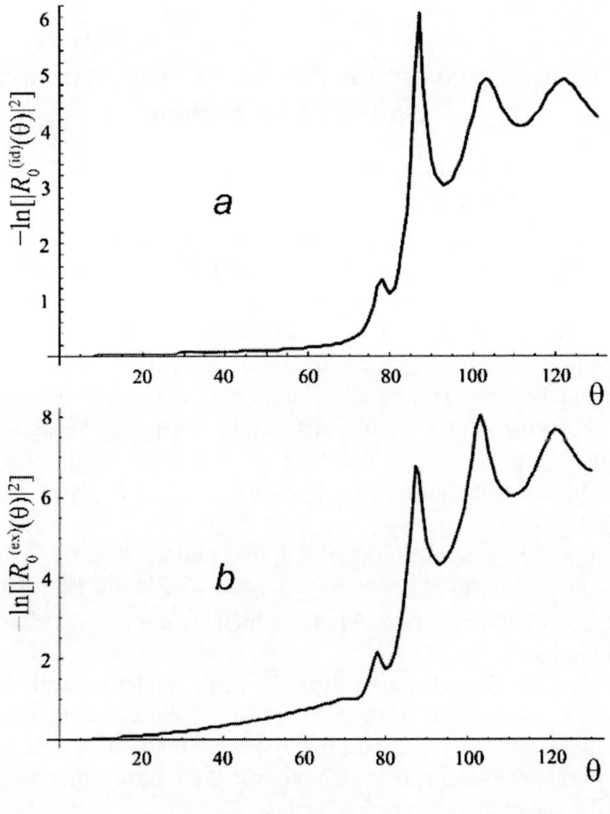

Figure 4.3. The simulated HRXR angular $-\ln[\ |R_0^{(id)}(\theta)|^2]$ - and $-\ln[\ |R_0^{(ex)}(\theta)|^2]$ -scans *versus* the θ-angle for the three-layer structure: $(a)\ \sigma_{n-1,n}^2 = 0$, $(b)\ \sigma_{n-1,n}^2 \neq 0$ (n = 1, 2, 3, 4).

Emphasize once more the idea of the NGD algorithm is to harmonize the process of the MLS vector **P** fitting. For this, *forcing* the Newton and GD algorithms to operate in alternative modes in an attempt to avoid undesirable problems associated with stagnation of the optimization process and the $F\{\mathbf{P}\}$-function slipping towards local minima. Such issues raise the most when the initial MLS fitting parameters are relatively far from the true ones *in search*.

In the case in question, the vector **P** rank equals 10.

The NGD algorithm code works determined by the relative number of successful events, which yield the true vector $\mathbf{P}^{(\text{true})}$, provided that the initial vector $\mathbf{P}^{(\text{start})}$ lies in a wide range around the true one.

4.4. Run-through for the three-layer structure. Results and discussion

In order to solve the inverse HRXR problem in the case of the imperfect three-layer structure, we use the simulated $|R_0^{(ex)}(\theta)|^2$-scan depicted in *Figure 4.3b*. The vector $\boldsymbol{P}^{(\text{true})}$ is defined by means of the MLS parameters as {{0.80, 0.90, 0.75}, {3.0, 4.0, 2.0}, {0.21 0.2285, 0.256, 0.2795}}. The dimensionless parameters are listed from the left to right as the sequential arrays, which contain the complex electric susceptibility $(1-0.05i)\{\chi^{(4-n)}\}$ for n = 1,2,3, the thickness $\{T^{(4-n)}\}$ for n = 1,2,3, and the mean-square roughness $\sigma^2_{4-n,\,5-n}$ for n = 1, 2, 3, 4, respectively.

The substrate complex electric susceptibility value $\chi^{(4)}$ is assumed to be equal to unity. The thickness values of $\{T^{(4-n)}\}$ are given in units of $10^{5/2}\lambda/2\pi$ the mean-square roughness values of $\sigma^2_{4-n,\,5-n}$ are given in units of $10^5(\lambda/2\pi)^2$.

The angle θ is given in units of 0.00039 radians, *i.e.*, θ_j, = $0.000039\,j$, where the integer number j runs from 1 up to J; J is the 1D rank of is the rank of the 1D angular array $\{\theta_j\}$, for which the input $|R_0^{(ex)}(\theta)|^2$ - scan has been evaluated.

To investigate the concurrent features of the NGD algorithm, for each event number *n,* the vector $\mathbf{P}^{(\text{start})}$ iteratively follows up to the different amplitude levels Λ = 0.1, 0.2, 0.3, 0.4; the iteration number k runs from 1 to 220. The number of events N =10. Note that if the error functional function value $\leq 10^{-9}$, the event is considered successful.

In *Figures 4.4(a)-4.4(d)*, the normalized factors $\overline{E}(\nu) = -\ln[F\{\boldsymbol{P}^{(k_{final},\nu)}\}/F\{\boldsymbol{P}^{(k_{start},\nu)}\}]$ *versus* the event number ν = n+1 are given (n=0, 1, 2, 3,…, 10). As it follows from *Figures 4.4(a)-4.4(d)*, the relative number of the successful events yields the probability estimate of the inverse HRXR problem solution with good accuracy when the true-valued vector $\mathbf{P}^{(true)}$ recovery is of the order of 90% - 40% for the values $\mathbf{P}^{(start)}$ with amplitude levels Λ from 0.1 up to 0.4, respectively

Based on the $F\{\mathbf{P}\}$ reduction trends in *Figures 4.4(a)-4.4(d),* one can claim that the NGD algorithm code provides a good vector **P** fitting to solving the inverse HRXR problem.

Even if the three-layer $\mathbf{P}^{(start)}$-vector lies far away from the $\mathbf{P}^{(true)}$-vector, the NGD algorithm successfully generates the $F\{\boldsymbol{P}\}$-processing.

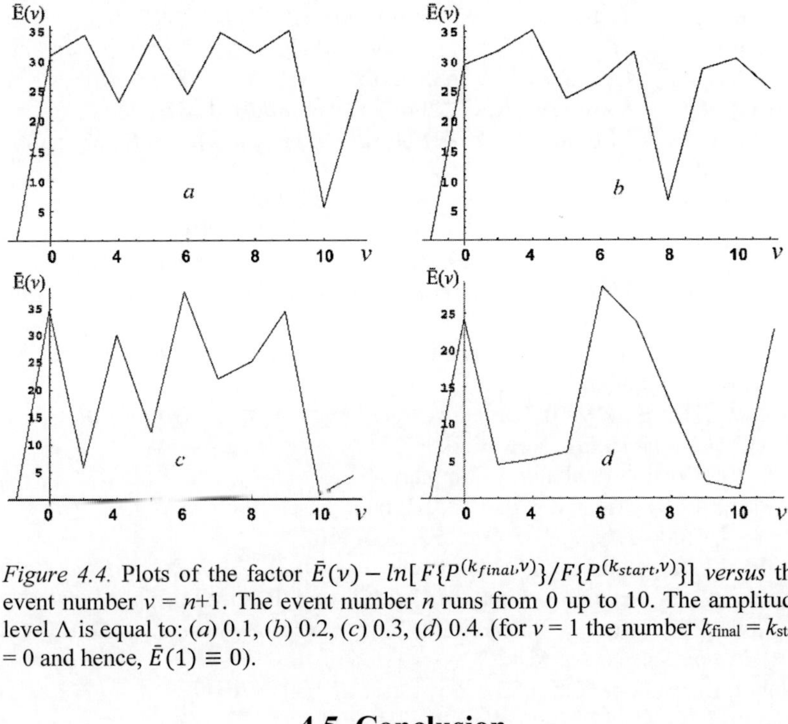

Figure 4.4. Plots of the factor $\bar{E}(\nu) - ln[F\{P^{(k_{final},\nu)}\}/F\{P^{(k_{start},\nu)}\}]$ *versus* the event number $\nu = n+1$. The event number n runs from 0 up to 10. The amplitude level Λ is equal to: (*a*) 0.1, (*b*) 0.2, (*c*) 0.3, (*d*) 0.4. (for $\nu = 1$ the number $k_{final} = k_{start} = 0$ and hence, $\bar{E}(1) \equiv 0$).

4.5. Conclusion

In concluding, one can state that the NGD algorithm implementation did successfully go through the test of recovering proper three-layer structure parameters. An essential feature of the NGD algorithm is the rapidity of unambiguous processing of the error functional function in a Chi-square sense to solve the inverse HRXR-MLS problem. In what is referred to as recovering the MLS data using the noisy HRXR angular scan data, the NGD algorithm application is still an open question.

At present, we can only state that the NGD algorithm is convergent and works well to decode the three-layer structure. How well it will work to solve the MLS data in a general case of the layer number more than three using the HRXR technique remains to be investigated. It is a good topic for a future job.

References

Baake, O., Oksuizoglu, R. M., Flege, S., Hoffmann, P S., Gottschalk, S., Fuess, H. & Ortner, H. M. (2006). *Mater. Charact.* ***57,*** 12-16.

Bridou, F., Gautier, J., Delmotte, F., Ravet, M.-F., Durand, O. & Modreanu, M. (2006). *Appl. Surf. Sci.* **253**, 12-16.

Bushuev, V A. & Sutyrin, A. G. (2000). *Poverkhnost,* **1**, 82-85.

Bushuev, V. A., Lomov, A. A. & Sutyrin, A. G. (2002). *Crystallogr. Rep.* **47**, 683-690.

Chukhovskii, F.N. (2009). *Acta Cryst.* **A65**, 39-45.

Cowley, R. A. & Ryan, T. W. (1987). *J. Phys. D Appl. Phys.* **20**, 61-67.

Eadie, W. T., Dryard, D., James, F. E., Roots, M. & Saboulet, B. (1971). *Statistical Methods in Experimental Physics.* Amsterdam: North-Holland.

Gill, P. E., Murray, W. & Wright, M. H. (1981). *Practical Optimization.* London: Academic Press.

Golberg, D. E. (1989). *Genetic Algorithms: Search, Optimization and Machine Learning.* New York:

Addison-Wesley Publishing Company, Inc.

Hodroj, A., Roussel, H., Crisci, A., Robaut, F., Gottlieb, U. & Deschanvres, J. L. (2006). *Appl. Surf. Sci.* **253**, 363-366.

Holy, V., Pietsch, U. & Baumbach, T. (1999). *High-Resolution X-ray Scattering from Thin Films and Multilayers.* Berlin: Springer Verlag.

Lemoine, P., Quinn, J. P., Maguire, P. D. & McLaughlin, J. A. D. (2006). *Carbon*, **44**, 2617-2624.

Nevot, L. & Croce, P. (1980). *Rev. Phys. Appl.* **15**, 761-779.

Parratt, L. G. (1954). *Phys. Rev.* **95**, 359-369.

Press, W. H., Flannery, B. P., Teukolsky, S. A. & Vetterling, W. T.

(1992). *Numerical Recipes in C: The Art of Scientific Computing.* Cambridge University Press.

Sinha, S. K., Sirota, E. B., Garoff, S. & Stanley, H. B. (1988). *Phys. Rev. B*, **38**, 2297-2311.

Chapter Five

The Electron Structural Crystallography

5.1. Introduction

Applying conventional direct methods and refinements for electronic structure determination raises some fundamental problems. There are some simple cases, *e.g.,* a surface, where such advances are legitimate as a good first approximation, but even here, correct results require the inclusion of dynamical effects (Marks & Landree, 1998). Theoretically, it is very well established that even one atom layer of a heavy element such as gold is a 'dynamical' scattering unit (see, *e.g*., Buseck *et al*., 1988; Spence, 1988). There are no cases with actual samples where the kinematical approximation is strictly valid. While exceedingly thin values such as 2 nm are sometimes reported for sample thickness used in interpreting experimental high-resolution transmission electron-microscopy (HRTEM) patterns, the actual thickness is undoubtedly larger. The discrepancy is often due to neglecting terms such as sample vibration or beam tilt. At the same time, under some optimal conditions, the HRTEM technique provides real-space information about crystal structures, although the HRTEM patterns generally do not contain actual atom positions.

On the other hand, to obtain the real-space information concerning atom positions, one may explore the direct methods relying on the diffraction information alone and extracting the unknown phase information among the structure factors (Dorset, 1995; Woolfson & Fan, 1995; Giacovazzo, 1998). Earlier, combinations of the HRTEM and direct methods have been tested neglecting dynamical effects in electron diffraction (ED) data, namely, the direct methods have been applied within a kinematical approach (Fan *et al*., 1991; Hu *et al*., 1991; Gilmore *et al*., 1993; Dorset, 1996; Marks *et al*., 1997). Noteworthy is that the first quantitative relationships used for phasing ED patterns *via* direct methods were the Sayre equation (Sayre, 1952; Dorset, 1995) and statistical phase invariants (Cochran & Woolfson, 1955; Hauptman, 1982).

Let us return to a set of structure factors {U(**g**)} (a set of the known moduli, {|U(**g**)|} is considered as the known; **g** is a reciprocal-lattice vector). The phase set {ψ(**g**)= Im[ln[U(**g**)]]} could be determined by using the iterative transform algorithm that is based on the Sayre-tangent equation, $\psi(\mathbf{g}) = \mathrm{Im}[\ln[\Sigma_{\mathbf{h}} U(\mathbf{h})U(\mathbf{g} - \mathbf{h})]]$ (Dorset, 1995; Giacovazzo, 1998; Landree *et al.*, 1997). The Sayre-tangent formula is relatively exact for point-like atomic structures as long as the atom scatterer positions are not overlapped, coinciding with the ED channeling approach (Van Dyck & Op de Beeck, 1996; Sinkler & Marks, 1999).

The phase information is preserved in the HREM patterns and can sometimes be extracted using the standard Fourier transform (DeRosier & Klug, 1968; Weirich *et al.*, 1996).

Ishizuka, Miyazaki & Uyeda (1982) have developed the phase-defined method proposed by Hoppe & Gassmann (1968). Combining the information in the electron micrograph and the ED patterns, within the scope of the weak-scattering approach, they have applied the method for phasing the structure factors of the crystal of copper perchlorophthalocyanine.

The least-squares refinement (Dorset & Gilmore, 2000) and direct methods (Weirich *et al.*, 2000) for decoding structural data from an ED experiment were reviewed in many publications, most of which ignored the dynamical ED effects. The direct method combined with the Fourier transform of the HRTEM pattern was used (Sinkler *et al.*, 1998) to solve the phase problem for the (Ga, In)$_2$SnO$_5$ ceramic.

Afterward, the two- and three-phase structure invariants, Σ_0 and Σ_2, within the scope of direct methods have been analyzed (Hu, Chukhovskii & Marks, 2000; Chukhovskii, Hu & Marks, 2001). In these works, direct approaches to the dynamical ED data can be understood *via* an 'effective kinematical approximation. Each of the phase conditional probability distributions, Σ_0, defined for many (**g**, **-g**)-reflection pairs, and the phase conditional probability distribution, Σ_2, designated for many (**g**, **h**, **-g -h**)-reflection triplets, were proved to display a strong peak in many cases.

The above assertion was discussed by Hauptman (1982). Mainly noteworthy is that the recovered effective dynamical potential may be similar to the kinematical one but does not have to be and, in general, will not be.

Hereafter, we push the direct method concept one step further using both the ED and HRTEM data sets as constraints for the phase-retrieval procedure. The domino phase-retrieval algorithm (DPRA) for direct phasing of the ED complex structure factors was proposed (Chukhovskii & Poliakov, 2003). It is based on the Gerchberg-Saxton algorithm (Gerchberg

& Saxton, 1972; Dainty & Fienup, 1987) and guides the relevant phase-retrieval procedure

A prerequisite for solving the phase problem was developing the computer convergent iterative algorithm, which can provide the phase problem solution in the electron structural crystallography. Using both the ED and HRTEM data, the DPRA has been approbated on the (Ga, In)2SnO5 ceramic example.

5.2. Phase problem formulation

Before formulating the domino phase-retrieval algorithm (DPRA) concept, we now use the necessary mathematical model describing the electron diffraction (ED) within the scope of 1s channeling approximation, where the electron wave function may be cast in the form (Van Dyck & Op de Beeck, 1996)

$$\Psi(\mathbf{r},z)=1+2i\sum_{j=1}^{J} a_j \Psi\left(\mathbf{r}-\mathbf{r}_j\right) e^{-i\pi\left(\frac{E_j}{2E_0}\right)kz} \sin\left[\pi\left(-\frac{E_j}{2E_0}\right)kz\right], \qquad (5.1)$$

where the sum is over the j-atomic positions in the two-dimensional unit cell, E_0 is the incident electron energy, the wavenumber $k = \lambda^{-1}$, the eigenvalue E_j refers to the lowest-lying 1s-eigenstate.

Equation (5.1) is the 1s channeling approximation for the wave function describing the electron propagation through the crystalline medium.

In the reciprocal space, the structure amplitude for the **g**-reflection has the form (Chukhovskii, Hu & Marks, 2001)

$$U(\mathbf{g}) = iF_{\mathbf{g}} = i\sum_{j=1}^{J} F_{j\mathbf{g}} e^{-2i\pi\mathbf{g}\cdot\mathbf{r}_j}, \qquad (5.2)$$

where the following notation is introduced

$$F_{j\mathbf{g}} = 2V_j(\mathbf{g}) e^{-i\pi\left(\frac{E_j}{2E_0}\right)kz} \sin\left[-\pi\left(\frac{E_j}{2E_0}\right)kz\right] \qquad (5.3)$$

for the complex atomic scattering amplitude of the j-atom, $\mathbf{r}_j$ is its position, and J is the number of atomic columns in the 2D unit cell.

Remarkably, the magnitude $V_j(\mathbf{g})$ is the Fourier transform of the 2D Eigenstate function $\Psi_j(\mathbf{r}\text{-}\mathbf{r}_j)$. In a first-order perturbation theory approximation, the $V_j(\mathbf{g})$ is the scattering factor of the jth atom.

The direct phase-retrieval problem is related to the complex structure factor set {U(**g**)} provided that the array {|U(**g**)|} is in hands. A {U(**g**)}-set forms the HRTEM pattern I(**x**) in the image plane of the electron microscope when the scattered electron beam propagates through the HRTEM *electron optical system* consisting of selected-area apertures and focusing lens. Along with this, the 2D array {|U(**g**)|} can be fixed *via* the diffracted intensity set in the back focal plane of the electron microscope.

The diffracted intensities are related to the ED set{|U(**g**)|}. In contrast, the I(**x**) is the modulus squared of the 2D Fourier transform of the complex structure factors {U(**g**)} with taking into account all likely aberrations, particularly defocus, spherical aberration, and so on. Thus, the 2D-periodical function I(**x**) is the HRTEM image of the crystal structure.

One can state that both the arrays of {|U(**g**)|} and {A(**x**)}, where $A(\boldsymbol{x}) = \sqrt{I(\boldsymbol{x})}$, can be considered as the permanent constraints in formulating the phase-retrieval issue underway.

From the mathematical viewpoint, each of arrays data {|U(**g**)|}, {A(**x**)} is non-convex. Accordingly, a non-convexity of both the above arrays is the main obstacle to applying the Gerchberg-Saxton-type algorithm (Gerchberg & Saxton, 1972; Dainty & Fienup, 1987), to solving the ED – HRTEM issue in question.

Then, it is physically reasonable to build the general phase-retrieval issue solution step by step, namely: beginning from the smallest rank ν of the arrays {|U(**g**)|} and {A(**x**)}, to their maximal *N*-rank feasible. In other words, to overcome non-convexity restrictions and avoid numerous redundant solutions, beyond the conventional Gerchberg-Saxton algorithm to phasing the whole system of **g**-reflections and instead apply the domino phase-retrieval algorithm (DPRA) for phasing structure factors in a sequential 'domino' fashion (Chukhovskii & Poliakov, 2003).

5.3. Domino phase-retrieval algorithm concept

Presumably, let us introduce the selected-area constraint sets of { $_{(\nu)}$|U(**g**)|}, $_{(\nu)}$A(**x**)} for the consecutive sets with ν = 1, 1+2, 1+4, 1+*N* (N is the even number, $U(\mathbf{g})|_{\mathbf{g}=0}$ is assumed to be unity for any ν).

For the selected-area array rank ν, the DPRA flow equation takes the form

$$
\begin{aligned}
&{}_{(\nu)}M_{k+1}(\mathbf{x}) = {}_{(\nu)}A(\mathbf{x})\exp[i\,{}_{(\nu)}\,\varphi_k(\mathbf{x})],\\
&{}_{(\nu)}|U_{k+1}(\mathbf{g})| = |InverseFourier[{}_{(\nu)}M_{k+1}(\mathbf{x})],|\\
&{}_{(\nu)}\psi_{k+1}(\mathbf{g}) = \mathrm{Im}[\ln[InverseFourier[{}_{(\nu)}M_{k+1}(\mathbf{x})]]],\\
&{}_{(\nu)}U_{k+1}(\mathbf{g}) = |{}_{(\nu)}U(\mathbf{g})|\,\exp[i{}_{(\nu)}\,\psi_{k+1}(\mathbf{g})],\\
&{}_{(\nu)}A_{k+1}(\mathbf{x}) = |Fourier[{}_{(\nu)}U_{k+1}(\mathbf{g})]|,\\
&{}_{(\nu)}\varphi_{k+1}(\mathbf{x}) = \mathrm{Im}[\ln[Fourier[{}_{(\nu)}U_{k+1}(\mathbf{g})]]].
\end{aligned}
\qquad (5.4)
$$

Each subsequent ED-HRTEM set includes the fixed complex values of $\{U(\mathbf{g})\,|, A(\mathbf{x})\}$-set iteratively obtained in all the previous ones.

For each *k*-iteration process within the cycle ν, the figures of merit (FOM) related to the reciprocal, $R^{(\nu)}{}_{k+1}$, and coordinate, $X^{(\nu)}{}_{k+1}$, spaces have been evaluated

$$R^{(\nu)}{}_{k+1} = \nu\textstyle\sum_{\mathbf{g}}\,|\,|_{(\nu)}U_{k+1}(\mathbf{g})\,| - |_{(\nu)}U(\mathbf{g})\,|\,|\,/[\sum_{\mathbf{g}}|_{(\nu)}U(\mathbf{g})\,|^2]^{1/2},$$

$$X^{(\nu)}{}_{k+1} = \nu\textstyle\sum_{\mathbf{x}}\,|_{(\nu)}A_{k+1}(\mathbf{x}) - {}_{(\nu)}A(\mathbf{x})|\,/[\sum_{\mathbf{x}}\,{}_{(\nu)}I(\mathbf{x})]^{1/2}. \qquad (5.5)$$

For reference, the phase set of ${}_{(\nu)}\{\psi_{k+1}(\mathbf{g})\}$ are sequentially calculated according to the flow-equation (5.5) for the numbers $k = 0, 1, 2, 3,.., K$. The terminated-procedure number K is governed by the FOM equations (5.5).

We want to make the following remark. To be specific, each phase set ${}_{(\nu)}\{\psi_k(\mathbf{g})\}$ embraces all the $\{\psi(\mathbf{g})\}$-phase values obtained in the preceding phase sets. The rest 'new' phases of ${}_{(\nu)}\{\psi\,{}_1\mathbf{g})\}$-set are chosen as random numbers in the range $[-\pi, \pi]$

Loosely speaking, the fundamental idea of the DPRA is the sequent phase fitting of the complex structure factors $\{U(\mathbf{g})\}$ step by step starting from the set ${}_{(\nu)}\{\,|U(\mathbf{g})\,|\}$, and ${}_{(\nu)}\{A(\mathbf{x})\}$ with $\nu=1$, ${}_{(1)}\{U(\mathbf{g})\,|\}=\delta_{\mathbf{g}0}$, $\delta_{\mathbf{g}0}$ is the Kronecker delta, $\delta_{\mathbf{g}0} = 1$ if $\mathbf{g}=0$ and $\delta_{\mathbf{g}0}= 0$ if $g\neq 0$. Then, following the flow equation (5.4), the Gerchberg - Saxon iteration algorithm is applied for all the subsequent cycles with $\nu = 1+2, 1+4,\ldots, 1+N$; N is the even number.

Further, for completeness, keeping in mind to control the numerical phase fitting process, is suitable to follow up the flow equation 'compliance' factor

$$C^{(\nu)}{}_{k+1} = 0.5\,\textstyle\sum'_{\mathbf{g}}\,|_{(\nu)}U_{k+1}(\mathbf{g}) - {}_{(\nu)}U(\mathbf{g})\,|\,/\sum'_{\mathbf{g}}\,|_{(\nu)}U(\mathbf{g})\,|. \qquad (5.5)$$

Here $C^{(\nu)}{}_{k+1}$ accompanies the *k*-iteration of the DPRA processing. The sum sign Σ′ is taken over all the **g**-reflections in the subsequent ν−cycles except $\mathbf{g} = 0$.

Notice that the redundant phase solutions can be readily detected by FOMs (5.5) when they take *inappropriate values*.

In principle, increasing the cycle number ν, the iteration number K to terminate the DPRA iterative process with the chosen FOMs-values enlarges.

Next, the DPRA is applied to phasing the structure factors for the $(Ga,In)_2Sn0_5$ ceramic using the input ED and HRTEM synthetic data.

5.4. Run-through phase retrieval for $(Ga,In)_2Sn0_5$ ceramic

The test case we attempt to resolve the phase problem is a center-symmetric structure of ceramic $(Ga,In)_2Sn0_5$. The structure is monoclinic with a = 1.169 nm, b = 0.317 nm, c = 1.073 nm, and β = 99.00°. The atom positions within the unit cell have been obtained (Sinkler et al., 1998). The projected 2D atom structure of $(Ga,In)_2Sn0_5$ is shown in *Figure 5.1.*

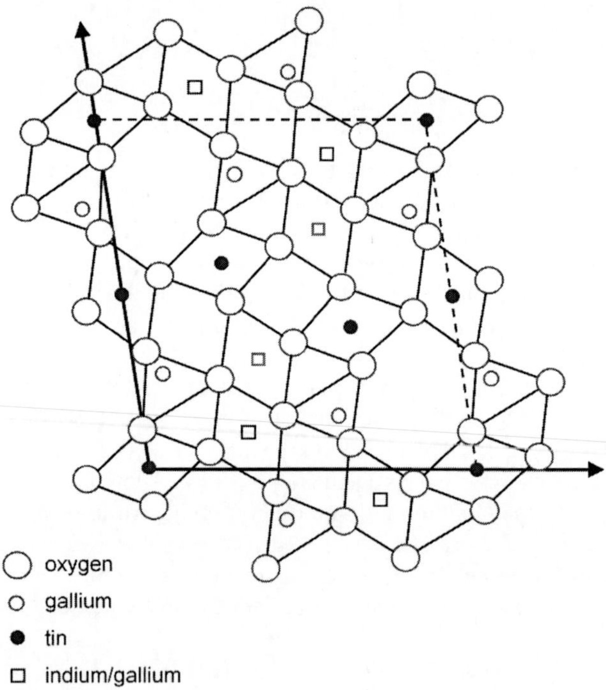

Figure 5.1. The 2D image pattern of $(Ga,In)_2Sn0_5$ was viewed along $[0\bar{1}0]$ (Chukhovskii & Poliakov, 2003).

The 2D and 3D (left and right) etalon HRTEM image patterns with a spatial resolution of 0.01 nm are shown in *Figure 5.2a* and *Figure 5.3a*, respectively.

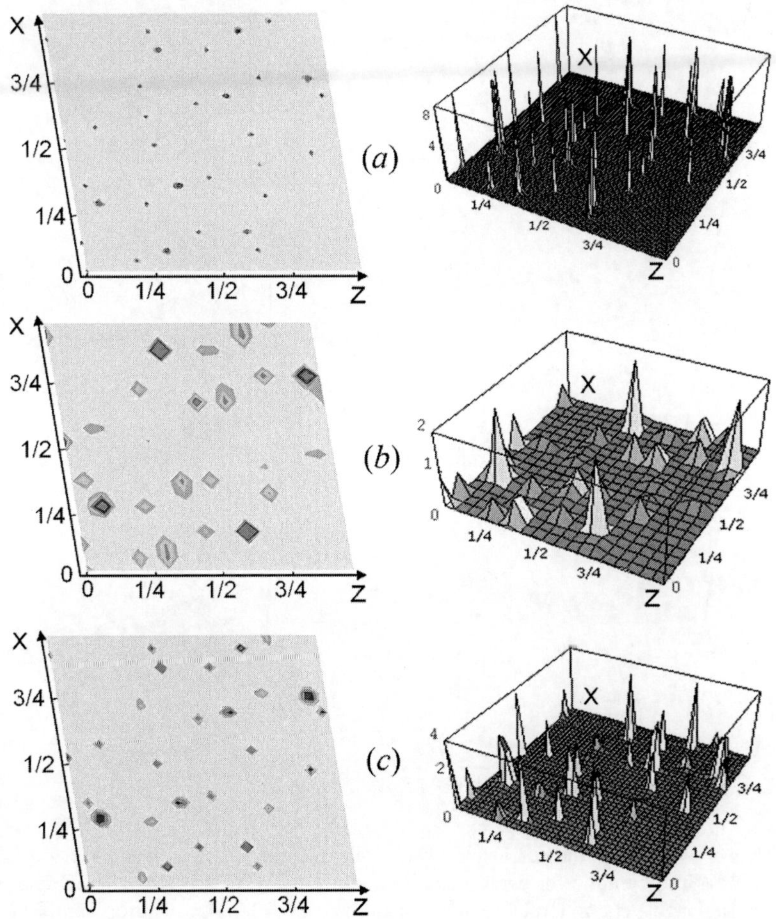

Figure 5.2. The calculated 2D and 3D (left and right) HRTEM image patterns of the $(Ga,In)_2Sn0_5$ ceramic for the cycle with N equal to (b)= 20, (c)= 40. The 2D and 3D (left and right) etalon HRTEM image patterns (a) = 100. The iteration number K is equal to 10^3. The sample thickness is equal to 4.121 nm.

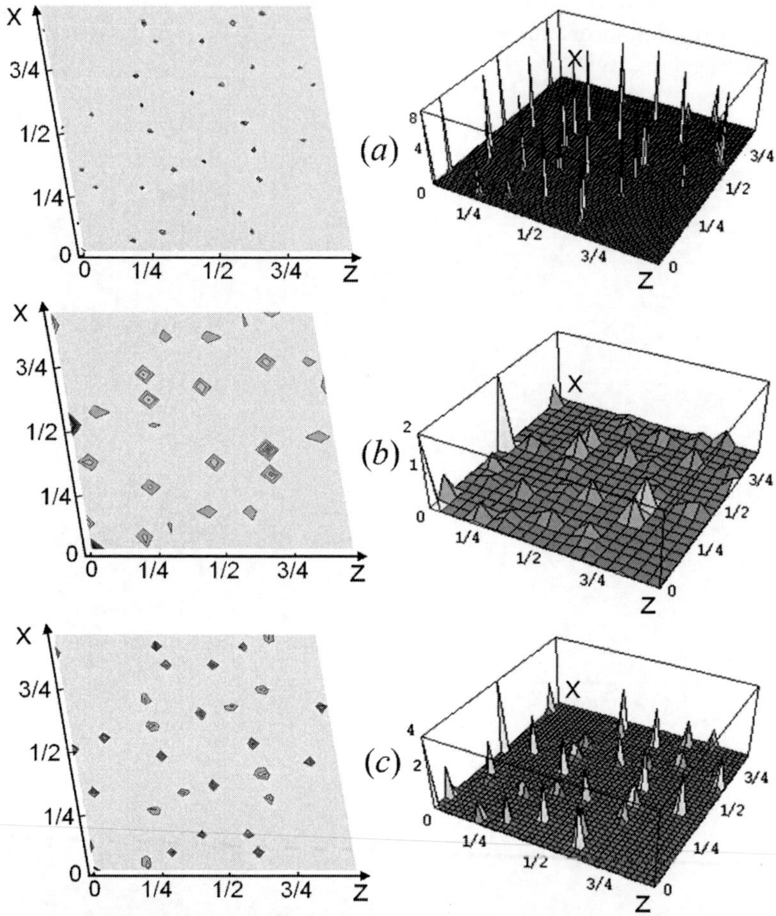

Figure 5.3. The calculated 2D and 3D (left and right) HRTEM image patterns of the $(Ga,In)_2Sn0_5$ ceramic for cycle with *N* equal to: (b)= 20, (c)= 40. The 2D and 3D (left and right) etalon HRTEM image patterns (a) = 100. The iteration number *K* is equal to 10^3. The sample thickness is equal to 7.291 nm.

To illustrate the DPRA code features that guide the phase fitting, the parametric plots of the *R*-FOMs (5.6) *vs.* the compliance *C*-factors (5.7) are given in *Figure 5.4*.

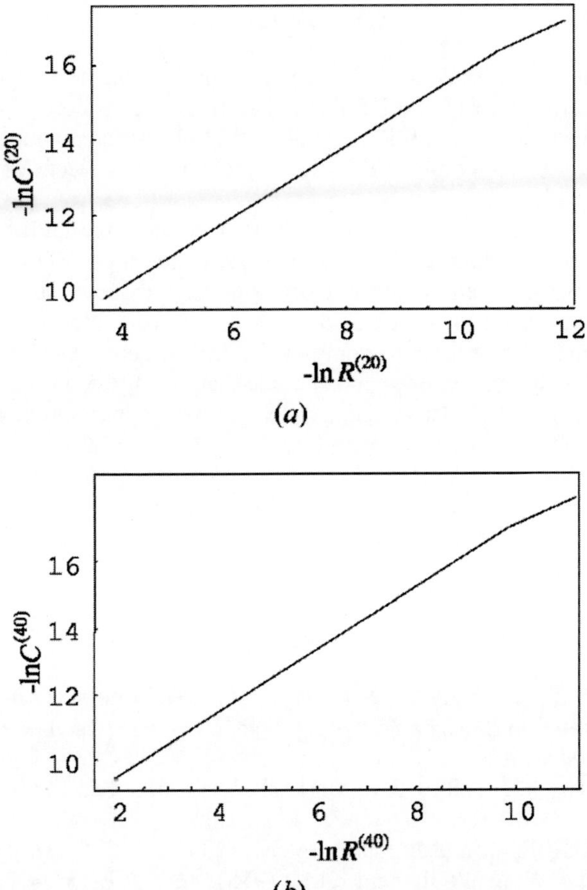

Figure 5.4. Plots of the factor $-\ln R$ *vs* the 'compliance' factor $-\ln C$. The even number N *is equal to*: (a)= 20, (b)= 40.Iteration number K = 100.The sample thickness of the $(Ga,In)_2Sn0_5$ ceramic is equal to 7.291nm.

From *Figure 5.4* it follows that for the calculated HRTEM image patterns displayed in *Figures 5.3b-5.3c*, the *R*-FOMs and *C*-factors achieve the values of the order of 10^{-5} and 10^{-7}.

5.5. Conclusion

A goal to justify using the DPRA for phasing the structure factors by using the artificial ED and HRTEM data may be considered as achieved. Noteworthy is that the DPRA provides the robust unambiguous iterative procedure for solving the phase problem in electron structural crystallography.

Decoding $(Ga,In)_2Sn0_5$ ceramic data shows how the DPRA algorithm works. Namely, using the large amount set of the input ED and HRTEM data, it works well independently on any extra assumptions and experimental conditions such as the electron weak-scattering and *a priori* structure models to phase the complex electron structure factors.

Based on our study, one can conclude that the DPRA above approbated is convergent and works well in application to the phase problem solution in electron structural crystallography.

Notice that when some {**g**}-reflections are not identified for rather complicated crystal structures, the latter can be feasible by extending the known set $\{_{(N)}|U(\mathbf{g})|\}$ by the extra {**g**-reflection} found due to the Sayre equation (Sayre, 1952; Dorset, 1995).

References

Buseck, P. R., Cowley, J. M. & Eyring, L. (1988). Editors. *High-Resolution Transmission Electron Microscopy and Associated Techniques.* Oxford University Press.

Chukhovskii, F. N., Hu, J. J. & Marks, L. D. (2001). *Acta Cryst.* **A57**, 231-239.

Chukhovskii, F. N. and Poliakov, A. M. (2003). *Acta Cryst.* **A59**, 48-53. Cochran, W. & Woolfson, M. M. (1955). *Acta Cryst.* 8, 1-12.

Dainty, J. & Fienup, J. (1987). *Image Recovery: Theory and Application*, edited by H. Stark. New York: Academic Press.

DeRosier, D. J. & Klug, A. (1968). *Nature* (*London*), **217**, 130-134.

Dorset, D. L. (1995). *Structural Electron Crystallography.* New York: Plenum Press.

Dorset, D. L. (1996). *Acta Cryst.* **B52**, 753-758.

Dorset, D. L. & Gilmore, C. J. (2000). *Acta Cryst.* A56, 62-67.

Fan, H.-F., Xiang, S. B., Li, F. H., Pan, Q., Uyeda, N. & Fujiyoshi, Y. (1991). *Ultramicroscopy*, **36**, 361-365.

Gerchberg, R. W. & Saxton, W. O. (1972). *Optik* (*Stuttgart*), 35, 237-244.

Giacovazzo, C. (1998). *Direct Phasing in Crystallography. Fundamentals and Applications*. Oxford University Press.

Gilmore, C. J., Shankland, K. & Bricogne, G. (1993). *Proc. R. Soc. London Ser. A*, **442**, 97-101.

Hauptman, H. (1982). *Acta Cryst.* **A38**, 632-641.

Hoppe, W. & Gassmann, J. (1968). *Acta Cryst. B24*, 97-107.

Hu, J. J., Chukhovskii, F. N. & Marks, L. D. (2000). *Acta Cryst.* **A56**,458-469.

Hu, J. J., Li, F. H. & Fan, H. F. (1991). *Ultramicroscopy*, **41**, 387-391.

Ishizuka, K., Miyazaki, M. & Uyeda, N. (1982). *Acta Cryst.* **A38**, 408-413.

Landree, E., Collazo-Davila, C. & Marks, L. D. (1997). *Acta Cryst.* **B53**, 916-922.

Marks, L. D. & Landree, E. (1998). *Acta Cryst.* **A54**, 296-305.

Marks, L. D., Plass, R. & Dorset, D. L. (1997). *Surf. Rev. Lett.* **4**, 1-4.

Sayre, D. (1952). *Acta Cryst.* *5,* 60-65.

Sinkler, W. & Marks, L. D. (1999). *Ultramicroscopy*, **75**, 251-268.

Sinkler, W., Marks, L. D., Edwards, D. D., Mason, T. O., Pöppelmeier, K. R., Hu, Z. & Jorgensen, J. D. (1998). *J. Solid State Chem.* **136**, 145-149.

Spence, J. C. H. (1988). *Experimental High-Resolution Electron Microscopy*. Oxford University Press.

Van Dyck, D. & Op de Beeck, M. (1996). *Ultramicroscopy*, **64**, 99-107.

Weirich, T. E., Ramlau, R., Simon, A., Hovmöller, S. & Zou, X. D. (1996). *Nature* (*London*), **382**, 144-146.

Weirich, T. E., Zou, X. D., Ramlau, R., Simon, A., Cascarano, G. L., Giacovazzo, C. & Hovmöller, S. (2000). *Acta Cryst.* **A56**, 29±35.

Woolfson, M. & Fan, H.-F. (1995). *Physical and Non-Physical Methods of Solving Crystal Structures.* Cambridge University Press.